DID THE SCIENCE WARS TAKE PLACE?

THE POLITICAL AND ETHICAL STAKES OF RADICAL REALISM

WILLIAM GILLIS

Everything in the universe is in the public domain, including this book.

Did the Science Wars Take Place ?:
The Ethical and Political Stakes of Radical Realism
First Edition, 2025

Published by Kindle Direct Publishing

ISBN: 979-8-218-75212-5

Special thanks to Frank Miroslav, Pete Wolfendale, Julianna Neuhouser, Nate Oseroff-Spicer, Scrappy Capy Distro, noBonzo.

Cover photo by Adam Jones. War Ruin alongside New Facade, Mostar, Bosnia and Herzegovina.

Center for a Stateless Society
c4ss.org

Subjects: LCSH: 1. Science—Philosophy. 2. Anarchism. 3. Discourse analysis

Contents

Foreword by Matilde Marcolli i

Introduction 1

Postmodernism as a Movement 11

The Metanarrative of Postmodernism 41

The War, The Hoax, and the Narrative 55

The Antirealist Constellation 71

The Stakes of Realism 121

The Antirealism Within the House 163

Just Found out About Object Permanence 171

Realism and Liberation 209

The Reactionary Rot in Postmodernism 225

Postmodernism's Influence on Reactionaries 237

Fundamentalism Among the Warriors 249

The Tepid Liberalism of It All 285

Conclusion 305

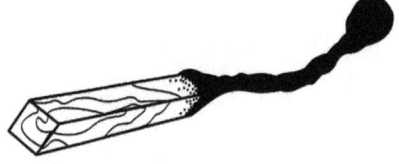

Foreword by Matilde Marcolli

This is a truly important and timely book. Fascism has come to power in the US. It now controls the largest nuclear arsenal in the world (a deeply ironic thought, considering how the official motivation for the original development of the atomic bomb in the Manhattan Project was ostensibly preventing such). It handles a technological apparatus of global surveillance never before seen, enhanced by the most advanced machine learning techniques sifting rapidly and efficiently through enormous amounts of data about the entire population, locating and targeting whichever class of individuals they wish to include in their lists of enemies of the state. It can unleash a massive propaganda machinery of automated production of disinformation through appropriately aligned AI generated text and images.

Yet, while brandishing some of the most advanced products of *technology* as weapons of violence and control, contemporary American fascism (with its political allies around the world) is also simultaneously engaging in an all-out war against *science*, from the gutting and dismantling of all the government agencies that throughout the postwar period supported the spectacular development of science, including the National Science Foundation and the National Institute of Health, to aggressive climate science denialism developed into an attempt to bring about the worst possible climate destruction (accelerated production and use of fossil fuels, abolition of all environmental regulations and testing systems, dismantling of any form of assistance for disaster victims and of the monitoring of risks and alert systems, such as hurricane and storm prediction at the National Oceanic and Atmospheric Administration). Medical research is being defunded to the point where famous cancer centers were forced to close down, an enormous number of clinical trials and treatment development programs canceled. Quackery is being aggressively pushed at the highest levels of government, amid the worse outbreaks of preventable infectious diseases seen in decades. Prestigious international medical journals are being threatened by the US government with aggressive demanding justification for their lack of inclusion of "competing viewpoints" not normally considered to be viable science. NSF grants are sifted through for "forbidden words" (marking scientific research as "neo-Marxist propaganda" for the use of such technical terms as "inequalities"), the entire directorate and programs structure of the NSF is being dismantled, entirely removing fundamental and basic scientific research, replacing technically competent scientific personnel with political appointees whose task is to declare scientific proposals politically acceptable.

Authoritarian regimes of the past have, of course, meddled with science in similar ways, from Stalin's science wars to Hitler's hunt for Jewish science, while still reaping the technological apparatus of modernity for belligerent use. Yet there is something that is especially prominent in current American fascism: extreme philosophical *antirealism*.

While anti-Rationalism has deep roots in the historical fascism of the past century, the absolute irrationalism, the blend of occultism, magical thinking, and the mysticism that permeated both Mussolini's and Hitler's fascisms, what is new in contemporary fascism is the way that *postmodernism* has served as an engine of legitimization.

This connection may seem at first very surprising, as postmodernism is typically associated with philosophical currents whose natural habitat were the corridors of the humanities departments of French and later American academic institutions during the 1980s and 1990s, and ideologically leaning towards various factions of the Left. How could this milieu give birth, during the following decades, to a fascist ideology embracing "alternative facts", denial of reality, and extreme cognitive relativism? Gillis' book provides a lucid and detailed analysis explaining this apparent contradiction, examining how the antirealist positions of Postmodernism were always in fact much more closely aligned with the ultranationalist idea of *different-cultures-producing-different-realities* promoted by historical fascism than with the internationalism and universalism of solidarity and freely shared knowledge foundational to socialism and anarchism.

Gillis traces the historical path that marked the spreading of postmodernist positions from French academia to the American left, between academic environments and activism. The collaboration of American science with military and capitalist enterprise alongside then gatekeeping and excluding from its ranks of large sectors of the population certainly contributed to the growth of anti-science sentiments in the political left, despite the prevalent view of science as radical, revolutionary, and emancipatory within the history of socialism. This friction, accompanied by the criticism and abandonment of the grand narratives of modernism, that increasingly shifted the focus on a fragmentation of perspectives, came to a clash in the '90s, with the so called "science wars" that Gillis' book analyzes in depth.

The events centered around a spoof article that radical leftist physicist Alan Sokal published in the postmodernist journal of cultural studies Social Text in 1996, as a way of criticizing the antirealist and relativist positions of the editors and their milieu. Variously misinterpreted, the original article and the controversy that followed marked a period of tense relations between different sectors of American academia and became largely forgotten some years down the line, the influence of postmodernism rapidly waning, while science was regaining popularity in the left as the looming climate crisis increasingly took center stage. Criticizing postmodernism started to look like beating a dead horse, until suddenly an aggressive wave of extreme right populism made its way to power in several countries brandishing a very similar brand of antirealism.

As fascism made a comeback as a global phenomenon, on a platform of ultranationalism and an anti-science agenda, the science wars of the '90s no longer look like a high brow debate internal to academic circles, but origins of increasingly existential threats.

Science denialism in the covid epidemic was responsible for a higher mortality rate that reached 43% after vaccine development. The US government's abandonment of international aid programs providing vaccinations and drugs for HIV, tuberculosis, and other conditions has been estimated by Nature as likely to cause the death of up to 25 million people within the next 15 years, arguably the largest genocide in history. Their aggressive climate disaster actions and inactions will also lead directly to large numbers of victims and massive displacements.

Alternative realities and denialism of the very existence of facts and verifiable hypotheses are not an academic debate: people's lives on a global scale are put at risk. Moreover, this is not a purely American phenomenon: Aleksandr Dugin, the ideologue of Putinist "national Bolshevik" fascism, continuing the legacy of Evola and esoteric Nazism, is mostly known within Russia as a postmodernist philosopher. Like American fascists, he scorns any notion of universal value or liberation, hiding his promotion of an aggressive nationalist fascism behind a curtain of cultural relativism.

All this while the occasional grifters posing as scientists on social media try to spin a fascist narrative around an alleged "support of science", which mostly boils down to pandering their own dubious or fringe theories, and muddling what would otherwise be very clear waters (a typical strategy of right-wing conspiracists).

Science is "radicalism," in the term's meaning of *getting at the roots of things*. It is also a critical instrument of anti-authoritarian resistance. Now, more than at any other time in history, the survival of the planet and of humanity is at risk. What liberation means today can only be global and based on universal shared values and knowledge: against the end-of-days death cult of fascism, science is our finest weapon and our greatest hope.

Matilde Marcolli is a professor of mathematics and computing and mathematical sciences at the California Institute of Technology (Caltech). She studied theoretical physics in Italy and mathematics at the University of Chicago and worked at the Massachusetts Institute of Technology (MIT), The Max Planck Institute of Mathematics, The University of Toronto, and the Perimeter Institute of Theoretical Physicists. She authored seven research monographs and over 150 scientific articles in mathematics, physics, information theory, and linguistics.

INTRODUCTION

Did the Science Wars Take Place?

In July of 2020, *Harper's* published a public letter decrying a supposed increasing wave of intolerance, public shaming, and moral certainty in society. While it exclusively namechecked Trump, the general intention and target was contextually clear upon release: it played into rampant tut-tutting among liberal commentators about "cancel culture" and activist militancy. The letter was signed by a hundred loosely "intellectual" figures of varying fame from David Frum to Cornel West, but most notably by the notorious transphobe JK Rowling, who had organized around it as implicitly a repudiation of the growing backlash she was receiving.

The signatories were a messy lot. Some, like trans activist Jennifer Finney Boylan, later protested that they had been duped by severe misrepresentations of context and Rowling's involvement, but it's fair to say for the most part they represented the sort of names a rich boomer with a deep allegiance to the liberal status quo would both recognize and think diverse.[1] The sort of emeritus professors and New York Times columnists that get namechecked by out-of-touch lawyers having dinner parties in brownstones.

Amid the riotous laughter from younger generations on Twitter, one leftist, Jaya Sundaresh, took it upon herself to make a thread working her way through the list of signatories, roasting each tepid liberal, dried out has-been, vapid grifter, and reactionary charlatan in turn.

That is, until she came across the name of her mother, Meera Nanda.

What followed went viral as Sundaresh live-tweeted her shock and apologia as she frantically tried to imagine how her radical leftist mother could have misread the context while trying to reach her. Seemingly the entirety of Leftist Twitter watched the mom and daughter drama play out live, and by the end of the day Nanda had stolen the best line of the entire affair: *"well if I'd known it'd get my daughter to call me twice in one day..."*

But let's be clear: duped or not, Nanda signed the letter.

For decades Nanda has faced threats as an intellectual critic of hindutva, the fascist movement that has held India in a chokehold since the mass-murdering pogromist Narendra Modi took power in 2014. But her specialization is as a philosopher of science and the attention she grabbed in the anglophone

---

[1] Amusingly, Glenn Greenwald was barred from signing it, showing that the drafters were conscious about their coalition and inclusions. They wanted to legitimize Rowling, not him.

world has been as a strident critic of *postmodernism*. Nanda was part of a large circle of leftists, anarchists, and physicists—most prominently including Alan Sokal and Noam Chomsky—that formed the other side of a very public and acrimonious conflict in the '80s and '90s called "the Science Wars."

Ostensibly a disagreement in academia that grew in rancor till it garnered front page newspaper headlines around the world, the Science Wars dragged philosophical questions to the center of political identity. Both the science warriors and the postmodernists agreed that "science" as a social matter warranted robust political criticism, particularly around funding, institutional structure, rhetorical use, and claims on biological essentialism or messy social matters. But sharp hostilities flew around matters of "realism."

> Do our models of the world really reflect it to any degree whatsoever? Is the project of science a struggle for a more accurate account of the regularities of the physical universe? Or is science closer to a purely social game where nothing beyond the social has any relevance or even an underlying reality? Is there a physical world at all?

Many of both the "science warriors" and the postmodernists used language from the radical Left and spoke of being motivated by liberation, but they split dramatically not just on this question of realism, but on what was at stake in it. The science warriors asserted that we can—however imperfectly, partially, and incrementally—grasp knowledge about an external physical world whose structures and universal regularities are not determined by our thoughts. Many postmodernists disagreed, on a variety of points. Both sides thought there were immensely pressing ethical and political stakes to the question, often believing that the other side was, beyond merely being ignorant or naive, functionally aligned with fascism or totalitarianism.

Even if the Science Wars have faded from direct attention or reference, the legacy of these public debates across the '80s and '90s continues to loom large. Their influence can be felt in almost every corner of today's political narratives, preoccupations, and enmities—from the Center to the Far Left and Far Right. But beyond the philosophical questions, even the bare account of *what actually happened* in and around the debate has been massaged and rewritten by a number of remaining camps until competing popular accounts differ so strongly as to be incommensurable.

Most notably, a host of reactionaries and bargain-basement "intellectual" grifters like Bret Weinstein and Claire Lehmann—loosely united in a coalition that goes by "The Intellectual Dark Web"—have attempted to steal the prestige of the '90s leftist "science warriors" while slurring pretty much every struggle for liberation as "postmodernist." Draped in the flags of "facts and reason" and "tolerant pluralistic liberalism," this menagerie of jumped-up podcasters has risen to such prominence as to have retroactively reshaped broad public awareness of the Science Wars and the points of contention.

Second to none in terms of said grifters is Jordan Peterson, a practitioner of the dying pseudoscience known as "psychoanalysis." Peterson is prone to rambling conceptual free association, slapdash genealogy, motte-and-bailey fallacies, Nietzsche fandom, explicit proclamations of epistemic relativism, fawnishly citing Thomas Kuhn, and even apologia for the nazi philosopher Martin Heidegger. And yet, he has—despite these postmodernist bonafides—somehow branded himself as a fervent *critic* of it. Peterson is infamously unconcerned with actually reading anything, but his main criticisms are obviously copied from Ayn Rand scholar Stephen Hicks' 2004 book *Explaining Postmodernism*. Hicks' thesis is basically that the Left reacted to the authoritarian failure that was the USSR by retreating to opportunistic sophism. This is not entirely devoid of historical substance and there are numerous books where leftists of varying ideological backgrounds level the same charges. But in Hick's right-wing framing, the central definition and sin of "postmodernism" becomes this *leftist* context and lineage—with liberal, centrist, and conservative postmodernists written out of the history. Thus in Peterson's hands the core problem with "*postmodern neomarxists*" is the *marxist* part (where "marxism" is treated so generally as to be synonymous with anyone holding egalitarian values or struggling for the liberation of oppressed groups).

This has reinforced a broad narrative among conservatives that goes something like this:

> Maybe once upon a time leftists played some small positive role in social progress, but they kept pushing more radical goals, and since their theories have no scientific grounding and their ideal of an egalitarian world is utterly impossible they have turned on anything to do with facts and logic rather than accept a moderate centrist compromise. So when you see strange radicals arguing for unfamiliar things—like saying the US was founded on slavey and genocide, or that gender is fluid and sex isn't a binary, or that it's possible to have a healthy society without a state or religion—rest assured that they only say these crazy things because they have entirely abandoned belief in or pursuit of truth for a sophism that valorizes collective self-delusion and has no compulsion against opportunistic lying. Their goal is to establish totalitarian control over society, where everyone is terrified to think differently or even assert that 2+2=4. This overall phenomenon is called "postmodernism" and you don't need to examine or engage with it. The absurdity of leftist claims is surely self-evident to you, and that is sufficient to prove they're all brain-poisoned zombies who are enemies of science.

With this narrative, any anarchist, leftist, or even just progressive liberal that says something beyond the immediate sensibilities of a conservative is immediately transmuted into a "postmodernist." And this has fueled an ecosystem of conservative grifters who frame themselves as frustrated liberals whose only allegiance is to "facts and logic," forced to side with conservatives against the frothing leftists and feminists corrupted by postmodernism. There's a killing to be made in telling 13-year-old suburban white boys that their initial perplexity at alien concepts like "white privilege" or "police abolition" or "patriarchy in video games" is totally correct by dunking on strawmen, shielding them from more involved arguments, and generally curtailing any intellectual inquisitiveness.

While this generates a dedicated fanbase that can be monetized, it has repulsed pretty much everyone else. The more these reactionaries whine about "postmodernism" the more that most people with a conscience or any ability to smell bullshit are going to be inclined to slide into self-identification with postmodernism or at least grow to instinctively dismiss any criticism of it.

Thus—better than any paid foil—the intellectual ineptitude and mustache-twirling misogyny of Peterson and his friends have polarized younger generations towards their bugaboo and accomplished a stark resurgence of postmodernism's standing from the disrepute it had increasingly fallen into by the close of the Science Wars. But these grifters have not accomplished this legitimization of postmodernism alone.

While Nanda may have herself been duped on the context of Rowling's letter, the track record of a number of other prominent leftist "science warriors" has not been great. To say nothing of his other crimes or embarrassments, Chomsky likewise signed the letter, and Sokal has explicitly sided with state repression of trans youth, even praising and helping edit the writing of the far-right grifter and conspiracy theorist James Lindsay.

Meanwhile, with their ignominy in the '00s dissipating and the political wind suddenly at their back, there's been a minor explosion of postmodernist media projects and online communities run by an archipelago of minor adjunct professors and failed grad students, excited to capture the allegiances of younger leftists and crush their rivals in old academic culture wars. These winds have buffeted a broader current in activist and radical subcultural spaces where deliberately irrational belief systems like astrology and chaos magick are praised as defensive alternate epistemologies of the oppressed.

Yet, in an era where the solidity of scientific facts like COVID-19 and anthropogenic global warming are simply not up for debate among young radicals, and where epistemic relativism has become the signature move of outright fascists, the stark antirealist or relativist positions of many postmodernists in the Science Wars present a liability. This has motivated the cultivation of some defensive narratives and reframings.

The current party line is basically:

> Postmodernism was never in any important sense a self-recognized body of people and never made any normative claims, it was simply the (neutral) diagnosis that the public had stopped believing in science and society had grown more complex and fractured. There wasn't anyone—certainly not anyone of note or influence—who was outright opposed to the hard sciences, endorsed mysticism, or encouraged epistemic relativism around things like gravity, evolution, viruses, and climate change. Postmodernists merely tried to raise the totally original point that social context and power systems can somewhat influence what gets established as scientific fact! But then some totally ignorant and unserious conservative physicists—who didn't like the leftist political implications, had never read the literature they were critiquing, and were looking to blame someone for their own funding getting cut—got mad that anyone would tread on their academic turf. The only thing they ever did, besides yelling insults, was exploit good faith to sneak a hoax article past peer review. Which obviously doesn't prove anything since journals in every field have lax oversight. This is so embarrassing for them that we should probably be nice and say nothing more about any of this (much less investigate it).

Pretty much every single aspect of this story is an audacious misrepresentation, but it's become ubiquitous in social media and in leftist subcultural spaces.[2] The extreme antirealist and relativist assertions pushed by many postmodernists in the '80s are obscured, watered down, or handwaved away. Their influence is denied, their critics are wildly strawmanned or selectively chosen and pretty much all of the actual history and context is avoided, as is any sincere coverage of stakes. One uses the sneery Jacques Derrida quote about it being *such a shame* that Sokal will be remembered for a hoax rather than real scientific work, and then end on a breathlessly fake-sincere urge for critics to do more reading. It's all gotten very paint-by-numbers.

The result of both the conservative and postmodernist narratives is to dissuade anyone from looking into the actual history and context of the science wars, as well as to collapse away the existence of the radicals and leftists who opposed postmodernism.

Many of the values and struggles that the reactionary grifters now slur as products of "postmodernism" were championed by anarchists long before it. While a divide within the Left between staid classical marxists and hip contemporary postmodernists is sometimes admitted to even by Peterson's ilk, it was neither dusty marxist economists nor avant-garde postmodernist

---

2 For a prominent compilation of all the usual bromides that was published when this book was nearly finished, see: "the physicist who tried to debunk postmodernism", *Dr. Fatima*, uploaded April 11, 2024, YouTube. https://www.youtube.com/watch?v=ESEFUaEA7kk.

zinesters that were pioneering the political positions and practices that most deeply horrify Peterson. They instead originate in no small part with feminists, antifascists, and anarchists far more radical than their postmodernist contemporaries, albeit usually outside of academia. Most of these very same radicals were furious critics of the postmodernist movement and the philosophical moves towards antirealism and epistemic relativism that infested it.

Both the conservatives and the academics share an incentive to hide all this. The academics tied to the remains of postmodernism seek to sanitize away its messy origins in subcultural spaces of the radical Left, because the philosophical arguments that energized their non-academic base in the '80s are embarrassing when aired directly and plainly. Similarly, the reactionary grifters are desperate to avoid direct confrontation with substantive arguments for things like transgender liberation and the obvious social construction of sex/gender—to give just one example—instead defensively asserting that relativist irrationality surely underpins every social justice struggle.

This cursed dichotomy, and the monopoly that both camps have on the discourse, has a feedbacking effect, pressuring everyone into functionally caucusing with one of these two camps. Even radical activists who oppose climate denial find themselves pulled into coalition with explicit and fervent irrationalists in hopes of together rhetorically resisting the reactionary scumbags wrapping their grotesque assertions in the mantle of "science." Meanwhile, the intellectual and infrastructural cores of the global fascist movement have explicitly embraced postmodernism, but this threat remains hidden to many on the Left because of the media prominence of the more mainstream conservative grifters.

These issues are all impossibly personal to me.

As an anarchist who came up in the radical left in the '90s and '00s, I'm intimately connected to a number of old comrades who now struggle with lifelong physical disabilities as a result of the batshit spiritualism and medical quackery frequently justified at the time with postmodernist arguments. On the ground, I've repeatedly watched as militant struggles were derailed by folks convinced that abandoning any notion of objective reality was a more profound strike against oppression, a strike against the hegemonic and imperialist metaphysical beast that is truth.

When liberal academics like Paul Feyerabend or Richard Rorty had the gall to appropriate our mantle and present themselves as "anarchists" or "anti-authoritarians"—to write entire books decrying science and realism on moral grounds as supposedly imperialist, totalitarian, and tyrannical—the results on actual anarchists and antiauthoritarians have been deleterious.

Additionally, like Meera Nanda, I grew up in a religious tradition very different from mainline christianity—one that rejects any objective reality of physical matter and that takes seriously the notion of plurality in reality construction. The result of this pluralism is wildly authoritarian and abusive, a fractal nightmare of constant gaslighting.

I have experienced the epistemic prescriptions of many postmodernists when taken seriously, and the result was not liberation.

At the same time, today I feel nothing but revulsion and betrayal at figures like Sokal who styled themselves as defenders of realism, but have since demonstrated very different priorities. The failures of many science warriors—their alliances with reactionaries and shocking deviations from anything like a radical realism—require at least as much careful autopsy and extraction.

This book is a cantankerous effort to resist the historical accounts pushed by conservatives and postmodernists of the science wars, trace their political influences and impacts, and highlight the radical realist position being obscured by both.

Chapter one, **Postmodernism as a Movement**, traces the cultural, political and discursive context of postmodernism's emergence in the late '70s, centering the story around *Semiotext(e)* in NYC and its many fractious relations with the anarchist movement.

Chapter two, **The Metanarrative of Postmodernism**, summarizes the content of postmodernism, as a rhetorical device, historical narrative, and set of values.

Chapter three, **The War, the Hoax, and the Narrative**, covers the infamous "hoax" of Alan Sokal, the immediate discursive context, and the attempted misrepresentations of it that have become commonplace.

Chapter four, **The Antirealist Constellation**, finally introduces the core philosophical issue of the Science Wars, laying out eight different notions of "antirealism" and citing many examples relevant in the Science Wars.

Chapter five, **The Stakes of Realism**, drills into the variety of sharp political and ethical implications that various factions in the Science Wars thought were at stake.

Chapter six, **The Antirealism Within the House**, talks about the role that certain physicists played in stoking and legitimizing antirealist claims.

Chapter seven, **Just Found out About Object Permanence**, defends a radical realism on grounds common among theoretical physicists, emphasizing the centrality of *reductionism*, properly defined.

Chapter eight, **Realism and Liberation**, discusses the broad conceptual and applied uses of realism for anarchists and liberatory social struggles more broadly.

Chapter nine, **The Reactionary Rot in Postmodernism**, exposes the extremely reactionary political sentiments and approaches that infested postmodernism.

Chapter ten, **Postmodernism's Influence on Reactionaries**, details all the ways that prominent fascists and other conservatives drew heavy and direct influences from postmodernism.

Chapter eleven, **Fundamentalism and the Warriors**, turns around and traces the reactionary political tendencies among the science warriors, drawing the knives of realism that have been sharpened over each prior chapter and using them to demonstrate how Sokal's attempts to defend laughably simplistic and transphobic accounts of "biological sex" clearly violate radical realism.

Chapter twelve, **The Tepid Liberalism of it All**, continues the autopsy of science warriors turned reactionary scum like Sokal, picking out how from the start their liberal political commitments prioritized civil discursive reason over any actual realism, much less radicalism, leading to inevitable reactionary capture.

And the conclusion draws it all together, summarizing the autopsy of both camps while emphasizing how those who share the radical approach to thought—in particular anarchists and physicists—can learn from one another. The necessity of a realist footing for anarchists, as well as the necessity of an insurgent footing—outside and in conflict with the state and political establishment—for scientists.

I wrote this book primarily to provide a map of the discourse for anarchists, feminists, and antifascists, who maybe didn't follow the Science Wars closely or came of age after it, but have nonetheless found themselves enmeshed in the endless fallout. But academics are technically allowed to read it too.

1

POSTMODERNISM AS A MOVEMENT

"Foucault's ideas, like Trotsky's, are never treated primarily as the product of a certain milieu, as something that emerged from thousands of conversations and arguments involving hundreds of people, but always as if they emerged from the genius of a single man."
David Graeber, *Fragments of an Anarchist Anthropology*, 2004

"I suppose that I diverge from "the methods of a philosopher" in this respect: that I am not interested in deconstructing and debunking the explicitly argued-for stances so much as the pervasive yet unspoken ones."
Mel Andrews, Twitter, 2024

"A story tells you how certain concepts arose, why they became important, why they changed and, above all, why they became a public malaise."
Paul Feyerabend, *The Tyranny Of Science*, 1992

To even speak of "postmodernism" is to immediately be subjected to a vast variety of denunciations and dismissals. Everyone has their own preferred set of authoritative texts or great authors, so any inclusion or omission is invalidating. Further, everyone has their own themes and issues they think should be the focus, with anything else mere irrelevancy and deflection. *It's a thin historical commonality! It's merely a gesture towards the zeitgeist of a past era! It's impossible to define! No one is defending postmodernism! No one identifies as postmodernist! Except for those that do, who all have completely different arguments and perspectives!* Because the term is often wide in who it catches, a great many people have developed a common investment in closing ranks to make sure nothing can be said about it as a whole. Even if this looks like a set of incompatible arguments that nevertheless interlock defensively.

To address postmodernism as a *movement*, a *milieu*, a discursive *community*, begins to capture this commonality, but this prompts new responses. *How can you meaningfully critique something as vaporous as a movement?! Are you really saying that a few humble little nothings could have an impact on anything? Surely, you know from your Marx that ideas are mere epiphenomena of historical forces!*

A very similar dance occurs with "neoliberalism," another movement of imprecise boundaries and frequent misrepresentation by critics, that often takes advantage of this to fervently deny its own existence, to say nothing of its history and crimes. Neoliberalism, we are told, was never a thing, whatever self-identified thinkers or arguments you might highlight don't really count as representative, and also it had no real impact on anything. *Look at how badly certain critics have critiqued it! Look at how bad those critics are! Surely you would want to be on the opposite side from them!* Perhaps it is not a coincidence that neoliberalism is likewise experiencing a rebirth from the ashes of ignominy, with increasing numbers of young people once again openly identifying with it, paralleled with astonishing chutzpah in historical revisionism.

To be clear: I am not totally without sympathy for postmodernists in this situation. There's something that feels starkly unfair about criticisms or evaluations across an entire *movement* when you, in all your unique specificity, belong to that movement. Yet such critiques, however necessarily rough or crude, are unavoidable if one wants to change the world.

When one critiques, for example, libertarianism, one inevitably has to address *common* arguments in the base rather than a few "canonical" texts, because Lysander Spooner, Ayn Rand, Robert Nozick, Rose Wilder Lane, Walter Block, Jo Jorgenson, Roderick Long, David Friedman, Michael Huemer, and so on all have very different positions, frameworks, and arguments. You will make absolutely no traction on critiquing libertarianism by narrow-beaming on any one of them or even a large set of them. The rank-and-file of the libertarian movement, rather infamously, *does not* align with the arguments of its leading theorists. It would be a grave mistake to exclusively argue with the writing of academic libertarians and never with the more popular arguments of the frothing reactionaries in the *Reason Magazine* comments section. No matter how closely a consensus in the movement might align with an argument in a specific text, it is not the same thing, and proponents of the popular variant will simply scoff and dismiss particularities of any text chosen to represent them. Even if you were to address every particular formulation of libertarianism, one by one, while you laboriously pressed a critique of any given position, rank-and-file libertarians would simply pivot back around to champion one of the positions you had already covered. This is how ideological movements protect themselves.

A movement's functional ideological content imperfectly maps to any set of Great Canonical Thought Leaders, because that is not how ideologies or movements work. There is never any single root to be found that can define the whole messy thing.

More importantly, the formalizations attempted by the famous academics are almost always a byproduct or regurgitation of wider notions and perspectives in the movement's base.

Anarchists are intimately familiar with this dynamic; our practices and arguments spread widely, but are never attributed in rarefied professional circles to activists but to some later random academic parroting them, whose class status alone magically transforms an idea into something respectably *citable*.

To be clear, this is not some complaint about ownership! Radicals and academics are playing two completely different games: radicals are usually happy to work anonymously so long as we can have an influence upon the world, whereas academia is structured around prestige games of citation that rarely prioritize broader impact. The ostensible parasitism of academics (and artists) who pick over radical subcultures and political movements to extract things of value is often more or less consensual. The problem is that the citation culture of academia skews how they view the history, development, circulation, and content of political movements.

If someone says, in critique of anarchism (as certain postmodernists have), that it has broadly embraced notions of an "essentially good human nature,"

I, as an anarchist, can whine and scream and thrash and point to exceptions or emphasize a different "core" of what counts as anarchism, but it would be almost totally beside the point to cite passages in the texts of Bakunin, Tucker, de Cleyre, Malalesta, Kropotkin, Bonanno, etc. that run counter to this. Even that diverse list of theorists are trivially not the same thing as anarchism.

If someone wishes to level such a critique at anarchism, they can and probably will cite books, yes, but at least as valid would be "*well I went to a lot of different Food Not Bombs and queer dance parties, read a lot of moldy zines or tumblr posts and have noticed some strongly recurring rhetorical patterns.*" And honestly, the moldy personal zines with print runs of 50 are probably better evidence than exegesis on Kropotkin's *Mutual Aid*.

Yes, arguments constructed with this sort of lens can be wrong, they can be skewed in the context and scope of their experience, but there is no other way to have an honest discussion about actual movements and actual ideological currents. After all, there are plenty of components of anarchism that are widespread and important without being ever written by some Thought Leader, perhaps even the *majority* of our theory that circulates as background consensus and deeply informs daily action has never been written down in such a form.

I have salted this book with footnotes, quotations, and references to pretentious Big Names, in part because I want to sketch an introductory map for those unfamiliar and help them draw out underlying connections, but I adopt this deplorable writing style in part to aggressively *mock* academic norms and highlight how weak they are at addressing anything important.

To restrict our attention exclusively to the Canonical Text or what academia considers sufficiently prestigious and permitted within the culture of citation is to simply protect ideologies and movements—ideas, as they actually course through the world—from all meaningful critique.

So before we turn to the heart of the Science Wars, it's important that we give an outline of postmodernism, *the movement*, and the social and discursive context in which its ideas congealed. Which, yes, includes some moldy zines.

French Student Fads

While use of the *term* itself goes back to uses from the late 1800s to early 1900s by different reactionary christian scholars forecasting an end of secular society and a return to faith,[3][4] when considering philosophy and politics (and so ignoring the somewhat distinct phenomenon of the same name in the art world), the *movement* of postmodernism has its early roots in the intellectual scene of post-WW2 France where structuralism was briefly quite popular.

This milieu had a shared background of conceptual frames and preoccupations typical of a significant chunk of Franco-German (so-called "continental") philosophy; they largely took for granted currents of moral and epistemic antirealism as well as those emphasizing supposed limits of rationality (in e.g. Kant and Nietzche) while taking seriously totalizing and sweeping accounts of society (in e.g. Hegel and Marx) that framed individual minds as mere products or parts of a wider whole.[5] But where figures like Marx had been ostensibly concerned with material context, these structuralists were more focused on language and how, as Lacan put it in *Desire And Its Interpretation*, "*a signifier does not concern a third thing that it supposedly represents but concerns another signifier which it is not.*"[6] That is to say, our words do not hang onto the external material world, but only refer to each other.

To many in this structuralist milieu, *even ideas* had no existence prior to language, but are inherently the products of the linguistic context we are born embedded within (no need to examine how prelinguistic infants think on their own, don't worry about it). Structuralism became so vacuous and mutable as a mass fad in France that by the '60s a trainer of the national soccer team declared they were undertaking a "structuralist" reorganization to win more games.[7] But in general it was just the latest variant of the same old anti-reductionist (e.g. anti-radical) impulse: emphasizing that you couldn't consider a given thing on its own, but had to rather view such in terms of the vast surrounding network of context, and in particular *language*—which, to be fair, is obviously true, in large part, for many things.

However one issue with attempting to view every specific thing in holistic

---

3 J. M. Thompson, "Post-Modernism". The Hibbert Journal XII no. 4 (1914): 733.
4 Bernard Bell Iddings, Postmodernism and Other Essays (1926).
5 See also how the philosophers of science they leaned most heavily on, Comte and Bachelard, spoke in terms of broad social stages or paradigms, applying holism not just over communities but eras.
6 Jacques Lacan, Desire and Its Interpretation: The Seminars of Jacques Lacan, trans Cormac Gallagh (1958), 11.
7 Francios Dosse, History of Structuralism: Volume 1, The Rising Sign, trans. Deborah Glassman (1997), 317.

terms of the aggregate social context it exists within, much less some perceived macroscopic structure, is that it involves extreme simplification: you're inclined to accounts of the whole that shrink away all possibility for change in the "parts" and miss how root dynamics may expose exceptions to macroscopic patterns.

Wherever you're talking about society or "a culture" or "capitalism" or "civilization" or whatever, holism has a tendency to collapse the trees into nothing more than *components* of a supposed forest. Every individual thing is taken as just a cog that reinforces the whole. And however hazy the concept of an individual tree might be, society-wide abstractions are innately vastly more hazy and arbitrary. The structuralists viewed themselves as rebelling against the notion that individual things have an innate nature to them—with specific concern about a supposed *human nature*. But they remained very much in the same vein as the marxists they rebelled against in making claims about the entirety of society, necessarily simplifying in the process and implying a staticity or solidity. Problem is, by the late 50s—after the crushed Hungarian revolution and Khrushchev's secret speech—the authoritarian instincts at the heart of marxism were increasingly starting to reek.

At this time the French intellectual scene had almost no contact with or real knowledge of the Logical Positivist tradition popular elsewhere and relatively little contact with the much wider analytic tradition in philosophy that defined itself not by responding to Hegel but by discarding him as an obvious hack. Even the most basic positivist texts weren't translated into French until the '80s, causing a late flurry of interest, and titanic philosophical figures in the rest of the world like Hayek and Rawls had the same delayed translation. And so this focus on grand holisms likewise meant they were still trapped in a very Hegelian dilemma: the question of how change or *resistance* can happen in a totalizing whole. Or, as it was rephrased, the question of *difference*.

In a network of signs that refer to each other rather than to an external reality, what gives "meaning" to one sign can arguably be its distinction from all the other signs. And if everything is merely a matter of such difference, well there's no interpretive finality, no final conclusive total picture to be reached. What young pseudo-intellectuals in Paris were hungering for was a reversion of the emotional associations in structuralism and an opening of possibilities beyond the static totality of stalinism.

> "The presence of an element is always a signifying and substitutive reference inscribed in a system of differences and the movement of a chain… this structuralist thematic of broken immediacy is therefore the saddened, negative, nostalgic, guilty, Rousseauistic side of the thinking of play whose other side would be the Nietzschean affirmation, that is the joyous affirmation of the play

of the world and of the innocence of becoming, the affirmation of a world of signs without fault, without truth, and without origin which is offered to an active interpretation."

Jacques Derrida, *Writing And Difference*, 1967

And so post-structuralism emerged as a new fad for a small number of pretentious Parisians, a language or framework that appeared to one-up the prior structuralism and could recognize some limits of it, albeit without actually breaking with the often implicit assumption that language and our very space of conceptualization is a self-referential game with no traction or reference beyond.

But it's important to note that few were really concerned with directly waging war on the idea of *a singular material reality* or the hard sciences. Even while some structuralists had started as materialists in the vein of marxism and, as post-structuralists, turned more anti-realist, the question of scientific realism vs. antirealism just wasn't a conscious or central issue. Anti-realist assumptions were either taken for granted or treated as valorously edgy, but physicalist or scientific realism—to say nothing of anti-realist but scientifically informed and preoccupied currents like logical positivism—weren't really in the room. Sciences like physics weren't a pressing discursive adversary and weren't being directly addressed or attacked.

> "if, concerning a science like theoretical physics or organic chemistry, one poses the problem of its relations with the political and economic structures of society, isn't one posing an excessively complicated question? Doesn't one set the threshold of possible explanations impossibly high?"
>
> Michel Foucault, interview, 1977 [8]

The phenomenologist Maurice Merleau-Ponty's attack on "*objective thought*" and "*unquestioned belief in the world*" or declaration that "*there is no world without a being in the world*"[9] was mostly background noise. The post-structuralists certainly had many edgy and arguably quite wrongheaded takes, but they weren't at this time focused on calling physics and mathematics nothing more than bad poetry with arrogant pretentions—as later postmodernists would. Their targets were the near-enemies of marxism, sociology, psychiatry, and existentialism—domains where their critiques often trivially landed.

What's critical is that while some post-structuralists touched Leftist currents, they themselves were not a *movement*, so much as a small avant-garde discursive community and a handful of celebrity figures, quite disparate in

---

8 Michel Foucault, "Truth and Power," in Power/Knowledge: Selected Interviews and Other Writings, 1972-1977, ed. Colin Gordon, trans. Colin Gordon, et al. (1988).
9 Maurice Merleau-Ponty, Phenomenology of Perception (1945).

interests and alliances. The central post-structuralist journal *Tel Quel* (named for Nietzsche's embrace of the world "*as it is*" and overseen by a reactionary catholic publishing house) was launched explicitly on the premise of rejecting politics for language and later denounced the 1968 student uprisings, with members sneeringly referring to some involved as "*anarchists*" and complaining about the "*stench*" of "*humanism*." While desperation to stay edgy or gain access to patronage networks saw this milieu flirt with stalinism and then an abstracted impression of maoism, such opportunistic political brandings never quite took. As a whole, the post-structuralist fad of a few snobby rich Parisian kids around inaccessible literary journals never managed to set roots down into the sincere Left or gain any popular traction and they collapsed in mainstream relevance in France as fashions moved on. By the '80s French academia at large had largely forgotten them and neoliberalism took over the public spotlight.

For postmodernism to become a real *movement* it had to be exported to America.

THE GENESIS IN AMERICAN RADICALISM

> "Perhaps this is a crude American misunderstanding of sublime and subtle Franco-Germanic Theory. If so, fine; whoever said understanding was needed to make use of an idea?"
>
> Peter Lamborn Wilson (as Hakim Bey), *T.A.Z.*, 1991

While French intellectual culture, by comparison, normalized a detached ironic playfulness and esoteric obscurantism, United States culture has long been overwhelmingly direct and *earnest*. Ideas are taken to *matter* here in a blunt way that often catalyzes dramatic shifts upon their import. In particular, the US has been significantly shaped by tens of millions of people taking philosophical idealism quite seriously.

"*Idealism*," in this philosophical sense of the term, doesn't mean its usual sense of having lofty ideals or values and pursuing them, rather it means the claim that either that our thoughts directly make reality or that our thoughts, in one sense or another, *are* reality.

Of course, the notion that matter doesn't really exist and the world is nothing more than a (collective) lucid dream is a very old and re-emergent perspective across human history. But it ran rampant across the 19th century US—new thought, theosophy, mesmerism, Christian Science, etc., etc.—these currents gathered immense followers and were significant presences in national discourse with headline legal trials and public intellectuals like Mark Twain feuding with them. There's an obvious resonance of *"simply believing hard enough will actualize something into being"* within populist narratives of The American Dream. While French catholicism has its marginal mystical and occult currents, by comparison the entirety of mainstream American culture was built on protestant notions of direct personal connections with the divine and a ravenous obligation to achieve individual success (as well as the notion that poverty or physical disability are the result of personal failings). Even the American pragmatist philosopher William James saw epistemic pluralism in terms of personal virtue, emphasizing a self-fulfilling role of belief, and positing the *"will to believe"* above the "absolutism" of science, justifying his own embrace of spiritualism, seances, and telepathy.[10]

The viral popularity of 19th century immaterialist spiritual movements never died out, but were reconfigured into large chunks of things like neopaganism, new age, prosperity gospel, etc. For the last forty years, cable preachers

---

10 William James, "The Will to Believe: Address to the Philosophical Clubs of Yale and Brown Universities" the New World, June, 1896. https://arquivo.pt/wayback/20090714151749/http://falcon.jmu.edu/~omearawm/ph101willtobelieve.html.

have stumped "*the law of attraction*" incessantly and books like *The Secret* can easily sell 30 million copies. Occult practices, seances, astrology, etc. have been in the highest echelons of American power, from Lincoln to Reagan. And the absolute apex figure in "*the power of positive thinking*," the clergyman Norman Vincent Peale, was Trump's beloved pastor and friends with Nixon. Hordes of capitalists are likewise enchanted by it; Steve Jobs infamously believed in his own "*reality distortion field*" and alternative medicine and so didn't get his cancer removed until it was too late, opting for acupuncture and psychic treatment.

In particular for our story, the chaos magician and NAMBLA ideologue Peter Lamborn Wilson (who wrote under the pseudonyms "Peter Cranston" and "Hakim Bey") heavily pulled from the american wingnut Noble Drew Ali's *Circle Seven Koran*, which, in turn, almost entirely plagiarized an earlier new thought text, merely replaced words like "god" with "allah" to give it an islamic aesthetic.

Because academia finds this stuff embarrassingly beneath notice, it is often written out of American intellectual history, treated at best as an amusing footnote concerning people who don't matter to serious intellectual issues, but discourses and movements are not confined to peer review, people have conversations constantly; they exist embedded in their surrounding culture. And in the late '70s, this broader constellation of immaterialist idealism and mysticism was overwhelmingly present throughout America, but definitely among college students, edgy subcultures, and the Left.

This is the context into which the remains of the post-structuralist French intellectual fad were imported, allowing for the genesis of postmodernism as an actual fervent *movement* with a social base.

Around this era, Foucault moved to San Francisco, made friends with analytic philosophers and denounced the French philosophical scene as mostly empty bullshiting. He was certainly not unknown when he arrived, his prior claims about the historical progression of notions of sexuality and criminality had obvious utility to queer activists and scholars and circulated widely. Meanwhile the rarefied aristocratic world of literary criticism (including many tweedy conservatives) were more than delighted to import and incorporate Derrida and Lacan. Gimmicks, like examining the signature of a book's author, were a novel flourish in literary essays, and the pseudoscience of psychoanalysis still garnered some respect in English departments, which were in the midst of their own wars over "new criticism." But none of this alone would have really made that big of a splash or congealed a *movement*.

From very early on, tiny artsy journals like *Diacritics* and *SubStance* in the US were reporting on *Tel Quel* events and mining French publications for snippets of decontextualized writing or theory they could republish that smacked of avant-garde prestige. But it took political countercultural circles to give postmodernism its footing. Eventually this would involve republications from publishing houses like Black and Red Press in Detroit (where the foundations of primitivism were brewing between the ex-marxists Fredy Perlman and John Zerzan). But nothing was more central in establishing ideological footing outside academia and creating "postmodernism" than *Semiotext(e)*.

In the early '70s, one of the marxist associates of *Tel Quel* at the Sorbonne, Sylvere Lotringer, by then a professor of French literature Columbia University in NYC, assembled his graduate students into a reading group and turned what they read into the journal *Semiotext(e)*.

Infamously, while he was working on the first issue in November of 1975, Lotringer invited Foucault, Lyotard, and Deleuze across the Atlantic to a conference he called "Schizo-culture." But to the horror of the French, this was less an academic or reputable conference than a bunch of unprestigious anarchist, feminist, activist, and countercultural rabble who had the temerity to do things like argue with them and denounce them, and so they offendedly abandoned the conference and retreated to hide at their hotel, avoiding Lotringer and turning down another invitation from him three years later.[11]

Despite their outrage at being insufficiently respected, Lotringer *was* a sincere fan. And despite *Semiotext(e)*'s deeply un-anarchist beginnings in academia, he was also involved in New York's subcultural milieus. His journal laudably rejected academic-style footnotes and was peppered with pornographic images and the sort of humor and aesthetics that typified the emerging zine culture. While Schizo-culture was a bust, by the '80s his friendships with two individuals in the subcultural world would pay off massively, building the traction and base for an ideological movement.

The first was Jim Flemming, of the Williamsburg anarchist publishing collective Autonomedia. And the second was previously mentioned NAMBLA mystic, Peter Lamborn Wilson, a former classics major at Lotringer's university, freshly back from serving as a family friend and court philosopher for the Shah of Iran, praising him in print as he slaughtered and tortured students.[12]

With these two friends and a small group of grad students, Lotringer began publishing the first translations of many core French post-structuralist books into English and marketing them to those in the local radical scene. Lamborn Wilson, serving as an editor of both Autonomedia and *Semiotext(e)*, would

---

11 François Cusset, French theory : how Foucault, Derrida, Deleuze, & Co. transformed the intellectual life of the United States, trans. Jeff Fort, Josephine Berganza and Marlon Jones (2008), 67-68.
12 Michael Muhammad Knight, William S. Burroughs vs. the Qur'an (2012).

begin referring to himself as a postmodernist or a "PoMo," and connections in the Libertarian Book Club in NYC plus the wide circulation of his zines as "Hakim Bey" would play the central role in spreading this obscure literary term from France as a political and subcultural identity.

Lotringer's dabbling with the non-academic world would continually cause him trouble, from anarchist and feminist denunciations of the "Loving Boys" pro-pedophilia issue of *Semiotext(e)* to his wife and collaborator Chris Kraus later facing uppity demonstrations from anarchists for being a landlord gentrifying New York. This hostile reception from the broader anarchist movement may be why, despite being distributed through Autonomedia, Lotringer consistently chose to publish marxists instead of anarchists, from the Red Army Faction to the Black Panthers. In 2000, *Semiotext(e)* would finally separate from Autonomedia, moving from the anarchist and punk scene to the graveyard of MIT Press. By the time Kraus drew fire for defending the postmodernist professor Avital Ronell's stalking and sexual assaults on a student, *Semiotext(e)* had long been pretty much dead to the anarchist movement.

Semiotext(e) was where postmodernism was actually born.

Its core base was the faddish avant-garde of trustfunders in the art scene and grad students or dropouts that weren't much involved in or connected to activism or militant struggles, but who still wanted to differentiate themselves from the boorish academia they were acculturated in. This meant a unique dynamic where they operated as rebel zinesters distroing what often amounted to edgy academic papers alongside collections of NAMBLA apologia and Nietzsche-obsessed chaos magick. Soon, beyond *Semiotext(e)*, you could find Foucault and Baudrillard repackaged and marketed in listings like *Sacred Jihad of Our Lady Of Chaos* with lines like "*the monolith of Consensus Reality is riddled with quantum-chaos cracks.*" Or drawing on Foucault, Derrida, and Feyerabend in Leonard Sweet's *Quantum Spirituality: A Postmodern Apologetic*. Similarly, in 1993, Out of Order Order (an occultist riff on Crowley) published the mystical and esoteric text *Liber AAA—The Art of Anarchic Artha*, citing postmodernism broadly as legitimization. Cross-publication was ubiquitous, to the point where the occult mystical screeds of Lamborn Wilson would be *explicitly* cited by some postmodernist academics as the genesis of their politics.[13] Through the wider zine culture's circulation, with varying levels of seriousness and wingnuttery, the academic remove broke down and imported fragments of the '60s French intellectual scene mutated into what actually found social traction.

The very popular writer on conspiracies and the occult, Robert Anton

---

13 Joseph Christian Greer, "Occult Origins: Hakim Bey's Ontological Post-Anarchism," Anarchist Developments in Cultural Studies 2 (2014): 168-187.

Wilson—a guest editor at *Semiotext(e)*—defined the emerging postmodernist movement in terms of *post-dogmatism* and rejection of "*mental imperialism*" (even while he ridiculed its social ties to academia and declarations that everything was socially constructed).[14] This sort of orientation and broad metanarrative was very popular in the '80s and there emerged a broad sense that "postmodernism" represented *just* that.

This built an actual self-reinforcing social milieu and identity, something far more stable and persistent than a momentary Parisian intellectual trend or academic research programme. Public attention could go away, funding could dry up, but a social movement with shared subcultural identity and a political movement who believes there are stark moral stakes at play won't all quickly pivot away when they can't find prestige in a poetry journal or an academic job. Indeed if someone becomes convinced an ideological analysis reveals something on which the entirety of social liberation hangs, they will remain a zealot until their death.

But academia was not irrelevant here; the solidification of postmodernism as an ideological movement with a subcultural base outside academia enabled a positive feedback loop with academic spaces. (A similar dynamic played out among the more transparent grifters of the "art world" although it had less impact.) Contact with activist movements as a source of novelty in concerns and ideas provided a great boon to what François Cusset derisively characterized as the "*theoretical careerism*" in postmodernism.[15] And many of the humanities were hungering for precisely what postmodernism offered.

At this time the anglophone academic world was deep in the midst of their own epochal saga, a grand manichean war that had been raging for decades, where the stakes couldn't have been higher: *funding*.

---

14 Robert Anton Wilson, Cosmic Trigger III (1995).
15 François Cusset, French Theory: How Foucault, Derrida, Deleuze, & Co. transformed the intellectual life of the United States, trans. Jeff Fort, Josphine Berganza and Marlon Jones (2008), 106.

The Two Cultures Cold War

In 1959, the chemist and novelist C. P. Snow gave an incredibly influential lecture called "The Two Cultures" that despaired at a widening discursive and conceptual gap between the sciences and the humanities. In particular, he zeroed in on the haughtiness increasingly prevalent in the humanities towards scientists, incorrectly assuming widespread illiteracy or ignorance of certain concepts or arguments, while simultaneously demonstrating egregious and shocking ignorance of basic scientific reasoning and knowledge. Snow lamented that the British educational system had for centuries aggressively overemphasized the aristocratic humanities (specifically Greek and Latin) and disparaged the sciences as low-class.

And this was certainly true. Before the second world war, the hard sciences carried relatively low-class associations and were hotbeds of leftism while in contrast the humanities broadly a place of retreat for the remaining aristocratic class or at least their culture and values. There were plenty of exceptions of course, not least of which the Royal Society, which infamously was a campaign to seize control over the diffusion of DIY journals between workers and craftsmen in which scientific research and results had previously circulated. But overall the hard sciences carried a particularly pedestrian cultural association closer to a trade school than a dinner party discussing Cicero. The mark of a classical aristocrat (like the brahmin in India's caste system) is not having to think about actual material reality; that's left to the help.

Moreover the hard sciences were clearly associated with radicalism—which is defined as attempting to grasp reality at the roots (*radix*)—whereas the humanities of the time were fixated on conserving culture and warning about the dangers of rapid change—the classic position of reaction. This was an incredibly pressing political polarization in the aftermath of the French Revolution, with reactionaries arguing that the radicals who thought they could understand enough about reality to improve society were dangerously arrogant.

As a result, prior to the second world war, the political valences in academia broadly placed the sciences with the Left and the humanities with the Right. There were exceptions, obviously, but the public associations at the time were skewed for good reasons. While historians today vary on estimates of e.g. the exact percentage of socialists and anarchists in the natural sciences, there's broad consensus that the skew was stark back then.[16] In the sciences you could find, for example, early inventors of cybernetics like Pitts and McCulloch

---

16 Eric Hobsbawm, "In the Era of Anti-fascism 1929-45" in How to Change the World: Reflections on Marx and Marxism (2012), 290-296.

living in a sexually libertine collective house while raising funds for anarchists fighting in Spain or Norbert Wiener's labor journalism. Theoretical physicists were seen as particularly far-left, with conservative politicians publicly decrying their "*abysmal ignorance*" of history and economics.[17] This was a context in which Albert Einstein's cousin fought for the anarchists in Spain, and Max Planck's son was executed for trying to assassinate Hitler. Meanwhile Oswald Spengler's widely read *Decline Of The West* denounced physics as a "faustian" corruption that was destroying life, and the Third Reich's education system focused on literature and elevated niches in the humanities like scandinavian history to the same funding levels as all of physics combined. This split practically paradigmatic: the cosmopolitan socialist Einstein vs. the return-to-nature nazi Heidegger.

> "The physicist by nature is politically radical. His mind is schooled in the proposition that progress is made by discarding various assumptions and premises and thereby making it possible to create a more powerful theory upon a simpler underpinning. The physicist, more than any scientist, deals with abstractions which make nonsense out of observations based upon the commonplace; he is educated in doubt and can disregard evidence which to the ordinary observer is both convincing and conclusive. Thus many physicists chose a vague leftist political philosophy."
>
> Richard Meier, The Origins of the Scientific Species, 1951

Pure scientists were heavily involved in antifascist and insurgent work across occupied Europe, holding a significant presence in groups like Groupe du musée de l'Homme and the Comité de vigilance des intellectuels antifascistes, doing things like organizing production lines for molotovs. Prominent figures like J. D. Bernal and J. B. S. Haldane were even regular streetfighters against Mosley's fascists in the UK.[18]

But the statist side of the war effort—from cypher breaking to the Manhattan project—compromised anglophone sciences dramatically, and afterwards the military industrial complex finalized control in the post-war glut of financing. Most radical leftists—even famous ones like Bohm—were systematically purged or marginalized as both inefficiencies and a threat to critical state security in both blocs of the Cold War. Theoretical physics in particular became an ideological enemy that was aggressively persecuted. David Kaiser details that more theoretical physicists were called as unfriendly witnesses

---

17 David Kaiser, "The Atomic Secret in Red Hands? American Suspicions of Theoretical Physicists During the Early Cold War," Representations 90, no. 1, (Spring 2005): 28-60.

18 Jedidiah Carlson, "Spread This Like Wildfire!," Science for the People Magazine, September 26, 2022. https://magazine.scienceforthepeople.org/online/spread-this-like-wildfire/.

before the House Un-American Activities Committee during its lengthy run than "*any other academic specialty: more than twice the number of historians… almost three times the nearly four times the number of economists or philosophers, and so on,*" as they were considered by endless reporters, judges and senators as "*inherently a breed apart, more susceptible to Communist influence than any other group of people.*"[19]

Beyond the aggressive blacklisting, the state endorsed the imposition of the "*Shut Up And Calculate*" approach, taught a cartoonish conveyor belt picture of a supposed "scientific method" in curriculums, and pushed for the eradication of longstanding philosophical discourse across physics as supposedly outside its domain.[20] Simultaneously, undeniable oppression of scientists by the USSR led to a mass exodus from the radical left by scientists horrified by the grotesqueries of Lysenkoism, etc.[21]

In some places like Italy, physics departments remained hotbeds of far-left radicalism, with figures like Franco Piperno and George Parisi playing huge roles. But inside the umbrella of the superpowers, radical physicists who maintained resistance against the military industrial complex and the politburo were consigned to activist projects like the Bulletin of Atomic Scientists, the International Committee Against Racism, or Science For The People, or hid in small circles in a couple cosmopolitan cities.[22]

On the other end, the sheer volume of students haphazardly pouring into post-war colleges and the system shock of Sputnik completely transformed US academia, providing unprecedented heaps of quickly allocated funding to universities which administrators redistributed relatively arbitrarily, enabling the creation or rapid expansion of departments and fields in the humanities (including sometimes the need for performative busywork), while creating an incredible anxiety over the risk of losing that same funding to the sciences.

Meanwhile, horror at Truman's use of the atomic bomb as well as the industrial tools applied in the holocaust sent shockwaves through the Left. At home in the US increasing conformity of technocratic corporatism and mass media helped build a general perception that the core of fascism had been homogenization and industrialism, and this was finalized in the attempt to define fascism in terms of a new concept called "totalitarianism" that paid little mind

---

19 David Kaiser, "The Atomic Secret in Red Hands? American Suspicions of Theoretical Physicists During the Early Cold War," Representation 90, no. 1, (Spring 2005): 28-60.
20 David Kaiser, "History: Shut up and calculate!," Nature 505 (2014): 153-155.
21 Gary Werskey, "The Lysenko Affair," in The Visible College: A collective biography of British scientists and socialists of the 1930s (1978), 292-304.
22 "Kaiser, David, "Nuclear Democracy: Political Engagement, Pedagogical Reform, and Particle Physics in Postwar America" Isis 93, no. 2 (2002):229–268.

to the ideology of Heidegger or Julius Evola as exemplars of fascism, to say nothing of the fascist *movement*. Instead, it presented fascism as the opposite of the liberal value of pluralism, diagnosing the core evil as essentially *universality*.

To the humanities partisans, the sciences became not just competitors for funding but the core of the monotonous life-draining megamachine that society had become. A great political repolarization was happening, where leftists increasingly found common ground with prior aristocratic currents in the humanities, seeing liberation in old tropes from romanticism and vitalism. This was the era of figures like Ivan Illich, Lewis Mumford, Norman Mailer, and—one of my father's faves—the marxist turned christian primitivist Jacques Ellul. Nowhere was this paradigm more omnipresent than in elite US schools like Harvard, where Ted Kaczynski was a then student, with his classmates later relaying how much his hyper-reactionary manifesto reflected the zeitgeist of their teachers.[23]

In the resulting atmosphere, marxists like Theodor Adorno and Max Horkheimer would shoot to wide circulation with tracts inveighing against "instrumental rationality"—a kind of narrow focus that takes ends for granted and seeks only formalized or quantifiable means—which they saw as typifying the sciences, supposedly indistinguishable from a wider technocratic apparatus. Science's radicalism (in other words its drive to find the root of all things) was framed as soulless and set in contrast to a vitalistic criticism that promised holistic analysis of social totality.

The first English translation of their *Dialectic of Enlightenment* came out in 1972, just as large swathes of the humanities were getting more confident about narratively situating themselves as the leftwing of academia. The deluge of students into the postwar college system had broken the back of the old guard in virtually every domain and much needed reforms and critical reappraisals were pushed everywhere. Since many fields like anthropology had been wildly racist, patriarchal, and colonial, it was easy and trivial to make vast advances in a short time, even if this also licensed overcorrections in the direction of "noble savages," gender essentialism with inverted values, or the crude anti-imperialism of campism.

The same wave of fresh blood in the post-war period also helped remake analytic philosophy, which had previously largely defined itself by rejection of idealists like the immaterialist George Berkeley, the crank Hegel, and the fascist Giovanni Gentile. Pre-war Anglo-American philosophy had *also* been broadly aligned with the technically anti-realist but nevertheless pro-science logical positivists, whose antifascism had forced them to flee Europe. But with

---

23 Alston Chase, Harvard and the Unabomber: The Education of an American Terrorist (2003).

the militant irrationalism of the Third Reich defeated, the urgency of these preoccupations started to decline. Moreover, as the two cultures polarized, analytic philosophy was awkwardly caught between the two, of no interest to technocracy (and hounded by the McCarthyist FBI), but out of step with the popular narratives of the humanities which greedily followed Adorno and Horkheimer in wildly misrepresenting the logical positivists both in philosophy and in politics.

Post-war analytic philosophers turned a variety of harsh critiques upon positivism, but they did so from many perspectives that would be unthinkably alien to Horkheimer's humanities partisans. Scientific realists critiqued positivism's crude anti-realist empiricism and pragmatics, while figures like Karl Popper engaged in a wider critique of verification; but all of these still were out of sync with the cultural repolarization.

> "1968 was a very revolutionary period, in many ways, the spirit of rebellion and revolt in many areas, and certainly also in philosophy… the logical positivist tradition was 50 years old. Well, at 50 it had to die."
>
> Bas van Fraassen, *The Semantic Approach to Science, After 50 Years*, 2014 [24]

It was Thomas Kuhn's sociological foray into the history of science that, while widely panned as unoriginal, unconvincing, and confused by reviewers in philosophy of science,[25] saw blockbuster interest (and, by his account, frequent *misuse*) from a younger generation in the wider humanities, eager to see scientific models reduced to arbitrary quirks of social history.

> "Twelve hundred persons were in the audience the night [the old view of science] died. It was March 26, 1969—opening night of the Illinois Symposium on the Structure of Scientific Theories."
>
> Frederick Suppe, *Understanding Scientific Theories*, 2000

Following that crossover success, came sharper provocations by figures like David Bloor, Paul Feyerabend, and Richard Rorty that leaned heavily on this audience and social base. There was no prestige to be won saying something boring and outgroup-y like *"the accounts of physics are broadly correct"* but you could easily pack lecture halls and sell out books raging against the "totalitarian" arrogance of science, or declaring that the *"tyranny of truth"* it

---

24 "The Semantic Approach to Science, After 50 Years," Princeton University and San Francisco State University, April 4, 2014, Rotman Institute of Philosophy, April 12, 2014, YouTube, 3:53. https://www.youtube.com/watch?v=6oM7-Wa_tAs.

25 Ziauddin Sardar, Postmodern Encounters: Thomas Kuhn And The Science Wars (2002).

represented *"could only lead to the most abject slavery."*[26] While critiques like Quine's "Two Dogmas Of Empiricism" could make a splash within academic philosophy, a mere couple decades later Feyerabend would publish *Against Method* not through a university press but through the state socialist publishing house Verso (then New Left Books), positively citing Lenin, Trotsky, and Mao throughout.

At the same time many who wanted to critique the sociological and institutional practice of science felt increasingly cocky that they could completely describe not just the social dynamics of scientists, but also explain away the *content* of scientific theories in terms of the social or historical pressures on the scientists. These academics chafed at the notion that physics or mathematics could provide strong boundary constraints upon their own studies as a more radically rooted field of knowledge, and they felt like philosophy had stolen explanatory accounts that were rightfully theirs. They struggled to create an alternative arrangement in which sociological knowledge was the basis of all other knowledge, giving sociology pride of place above physics. The extremists among these ideologues sought to reverse the foundations and causality, arguing that the regularities studied by hard sciences were phantoms determined entirely by the contingencies of the social.

It's easy to see the political pressures behind this move, in a context when capitalists and racists would regularly retreat to making claims in the trappings of science so as to position them beyond social critique. A theory of population genetics, for example, is clearly influenced by social context in terms of motivation and institutional inclination, and it's quite reasonable to examine how social matters can influence the content of such theories. But the pugnacious inclination was to go further, to declare not just that the social can, to some measure, influence the content of scientific theories—this was never really in question—but to also declare that the social constitutes *the only influence*, that any voice of a material reality beyond the arbitrary happenstance of human social conditions was not present or irrelevant.

In this context, the congealing postmodernist political and subcultural movement constituted a base mobilized externally to academia, but also close enough to apply pressure within it. And it provided an ideological banner that could push the envelope of narratives already in vogue across the humanities. From within academia, to associate yourself with postmodernism was to instantaneously acquire a broad pool of supporters and an image of bravely taking the next logical steps in what was a widely popular consensus.

While the journal *Social Text* would rise to central notoriety in the Science Wars, its origins are particularly illustrative. When it launched in 1979, it did

---

26 Paul Feyerabend, The Tyranny Of Science (1996), 74.

so as an avowedly marxist journal with broad scope across the humanities and social science. It namechecked Derrida and deconstruction but also Chomsky and his linguistics program, treating both as merely two leftist academic projects of interest. There were hardly any discussions of natural science, much less rejections of physical realism, and an article on the coup in Chile even made, in passing, an explicit distinction between science and *corrupting* ideological influences on it. At this inception there was not much focus on poststructuralism and little evident awareness of the postmodernist milieu Lotringer was congealing with *Semiotext(e)*. *Social Text* published the occasional pseudoscientific wingnut who was into stuff like "orgonomy" but that was it.

Founding editor Stanley Aronowitz was ideologically representative of the academic left at this moment. Aronowitz had come up as a stalinist in the CPUSA and migrated to the New Left (he was one of the authors of the Port Huron Statement). While working as a professor in California, he became friends with Herbert Marcuse and emerged as a plumbline ideologue of the Frankfurt School, embracing Adorno and Horkheimer's metanarrative where the central villains were instrumental reason and "positivism." In this respect, he was largely indistinguishable from the academic left of the time—holding a few wingnut sympathies and prone to rants about how the hard sciences were winning funding over the humanities. But, after he moved back to New York City in 1982, Aronowitz found the postmodernist movement growing at the intersection of academia and radical spaces and became one of its most bombastic and pugilistic champions—even more brash than the other academics feeding on the avant-garde political fringe and more than happy to say certain quiet parts aloud. While Lamborn Wilson lurked around occult circles for failed beatniks, Aronowitz was interested in wider audiences, repeatedly speaking on public television and helping to found the DSA. Where others preferred obscurantism and ironic feints in closed quarters, he took their claims earnestly.

By 1988 Aronowitz had written the jeremiad *Science As Power*, a book that attempted a grand move beyond his Frankfurt School influences, integrating the burgeoning postmodernist milieu and the sociologists' attack on science into a defense of marxism. Where Marcuse saw science as *capable* of producing accurate models of reality separable from the historical context in which they were produced, Aronowitz embraced a sharper relativism, seeing science as entirely the ideological product of the institutions of capitalism, even confidently predicting that science and the state would merge entirely, with technocracy replacing all other ideologies of power. Science, in his account, was merely an unjustly hegemonic narrative, with no solid claim or access to truth in comparison to things like creationism.

The political motivations behind this were explicit. Aronowitz had correctly perceived that the inclinations of the Frankfurt School (and marxism more generally) towards "holism" and "dialectics" were existentially challenged by physics' radical focus on underlying root dynamics. And so, to salvage the marxist project and the humanities more broadly, it was necessary to throw off Adorno and Horkheimer's residual sympathies for science and attack it in its entirety.

While the book was unoriginal, mostly consisting of Aronowitz summarizing (often poorly) a number of arguments and perspectives across the academic and radical Left—imperiously framing each as an authoritative advance—it's an excellent example of the cross-fertilization that was catalyzing. Among the summaries of various arguments, he interjects a number of complaints against the common belief that causality flows one way, that biology is explained in terms of underlying chemistry, and "*the denial of intentionality in nature*" [27] while gleefully trumpeting the supposed triumphant reemergence of wingnuttery from *teleology* to Lamarckianism. The book is riddled with pompous misunderstandings of physics, especially around relativity and quantum mechanics, like that it refutes all logic and "destabilizes" mathematics—interpretations which it triumphantly leans on as justification for extreme epistemic relativism. The attempted conjunction between such tortured misunderstanding of physics concepts and his impressions of European poststructuralists was quite explicit:

> "Recent social theory in France, Italy and Germany roughly follows the prescriptions of relativity and quantum theory in their respective insistences on temporal discontinuities, spatial indeterminacy, and historical construction of discursive formations in social phenomena."
>
> Stanley Aronowitz, *Science As Power*, 1988

---

27 Stanley Aronowitz, Science as Power: Discourse and Ideology in Modern Society (1988), 246.

ANTIREALISM TAKES OFF

Social Text and Aronowitz were just one part of a wider emerging scene that hybridized between radical subcultural circles and academic circles. For example, in 1996—a year that will be important in our story—Aronowitz would edit, publish and praise Lamborn Wilson in the compilation *Technoscience and Cyberculture*, with a piece where Wilson rambled at length in apologia for pederasty, child rape, and more general violations of consent, explicitly drawing upon Feyerabend for legitimization.

> "If we must be crude about it, we shall have to declare *in favor* of "boundary violations." ... The real doomsayers are the proponents of order and progress, whose world view reduces them to a hysterisis of rigidity and body-slander. But the proponents of a Feyerabendian "chaos" (an *antitheory*) are in fact the true biophiles, the party of celebration."
> Peter Lamborn Wilson (as Hakim Bey), *Boundary Violations*, 1996

One could question whether Feyerabend would directly cosign Wilson's underlying occult framework in which reality was a formless chaos devoid of any order, but the dadaist was freshly dead and thus even more easily mobilized by the NAMBLA theorist Wilson, who in turn was happily mobilized by the marxist Aronowitz.

In some ways this postmodernist scene replicated the French poststructuralist milieu of *Tel Quel* that it pulled from—Parisian youth in the '60s had been swept by esoteric obsessions with conspiracies and magic, with books like *The Morning of the Magicians* becoming wildly popular—but there were significant differences. Anti-realist moves that in France had been coquettishly esoteric, the American masses embraced ever more bluntly and fervently.

Significant translation jumps with figures in circulation in Europe expanded and redirected critiques—a German term like "Wissenschaft" that means something looser than a body of knowledge and may connote something closer to theology and art in a context can become translated as "science" in English and then greedily directed in critique towards the STEM barbarians. Obscurantist elitism mixed with zealous populism and inter-departmental scuffles in often complex feedback loops.

And as the waves across academic spaces grew, they prompted backlashes that bolstered their own standing. The more conservatives screamed about postmodernism eroding the sanctity of the western canon in literature, the more postmodernism as a whole gained prestige with the wider Left. Even if folks thought certain declarations among their ranks silly or unserious, members of the movement had no motivation to object to increasingly strident provocations, but they could unquestionably delight in the howls of objectors providing them all free advertising. It's a consistent dynamic of edgelord schismogenesis: The more something pisses off the enemy camp, the more it is given a pass, the more internal challenges to it become unthinkable treason, and the more it becomes synonymous with your broader position.

Such moves within western academia were paralleled and reinforced by a wave of decolonial theorizing in the global south that emerged to explain why the success of nationalist movements had not secured liberation.

To give one example, in Ghana upon independence, the marxist Kwame Nkrumah had established a brutal dictatorship and outlawed labor strikes, pillaging the people to enrich his immediate circles. In one instance, 80,000 individuals across 700 villages were violently displaced to create the Akosombo Dam, whose electricity went to aluminum smelting arranged on generous terms with US capitalists, before Nkrumah recanted under outcry and nationalized the economy into a mismanaged failure. To redirect the anger of those he had oppressed, Nkrumah wrote a book that would go on to influence generations of theorists, *Neo-colonialism: The Last Stage of Imperialism.*

Instead of tackling the new hierarchies of governments and local elites, the violent divisions of class, borders, and national identity (as with Nkrumah's ethnic cleansing of the Yoruba), or the continued projections of force by the old colonial masters in military entanglements and predatory loans, the theorists of neocolonialism tasked themselves with finding less challenging enemies. Struggle for tangible material freedoms was dangerous to the new regimes, far safer to redirect the hunger of the oppressed into the fetishization of *local identity* in a forever war with foreignness. It was not long in these spaces before science was elevated into an external imposition corroding native culture, even said to be the core blueprint of imperialism.

> "There must be skepticism against science [because] modern science is the basic model of domination of our times and is the ultimate justification for all institutionalized violence."
> Ashis Nandy, *Science Hegemony and Violence*, 1988

To be clear, it goes without saying that scientific practice has been enmeshed in mutual influence with colonial or imperialist regimes.

Throughout history, whenever scientific thinking has developed broad respect and cache, authorities have appropriated its aesthetics to drape over their own undertakings and attempted to buy, control, or otherwise influence scientists. Like anything else, the social landscape of scientific practice is a medium through which power can exert itself. One can laugh about how phrenology was even in its era recognized as obviously fake and a blatant attempt by grifters to steal the prestige of physics, but actual scientists have done immense harm via integration with ongoing colonial projects in a great multitude of ways. There's nothing inherently wrong with fossil hunting to build a better picture of theropod diversification in the Cretaceous, but a British paleontologist was at best going to get permission from his imperial compatriots and callously benefit from the violence the Empire could project, not collaborate in a truly egalitarian fashion with locals. Similarly, rich western scientists *still* often parachute into the global south to extract data and leave, perhaps paying a few fixers a pittance but ignoring both the insights of local communities and local scientists while leaving little to no knowledge, funding, prestige, or lasting connections. This entire book is not enough space to list diverse examples of how the daily practice of science can often grow in alignment with and reinforce the interests of power. (And similar stories could be told about the applied practice of even *basketball* or *music* and their intermingling with the power structures of our world.)

But the critiques increasingly being pushed were not of such dynamics, but of what they said constituted the modern materialist "scientific worldview" at core:

> "Capitalism is based upon a cosmology that structurally dichotomizes reality: the one always considered superior, always thriving, and progressing at the expense of the other. Thus nature is subordinated to man, women to men, consumption to production and the local to the global."
>
> Maria Mies and Vandana Shiva, *Ecofeminism*, 1993

In Vandana Shiva's extremely influential telling, before colonization India was a relatively harmonious feminine society, closer to unity with nature. Everything changed when science arrived and imposed its dualistic alienation from nature, creating masculine domination over the earth.

This is, of course, an audacious and immediately apparent lie to anyone other than guileless white hippies. Pre-colonial India was ridiculously hierarchical and patriarchal, with strong hostilities to nature or matter as inferior and polluted. This is no small part of why materialist and atheist currents in India were long associated with the oppressed.

> "The contradiction between the philosophical idealism of Vedanta and the philosophical materialism of the heterodox, non-Brahminical castes who practically made up the entire class of workers—peasants and craftsmen—who actually came in contact with the elements of nature."
>
> Meera Nanda, *Prophets Facing Backward*, 2004

While growth in scientific knowledge certainly expanded the *scope* of material projects that could be feasibly implemented by authorities (precolonial, colonial, and postcolonial), contrary to Shiva, the more refined strategies of investigation of modern global science never constituted some total break, nor was the "scientific worldview" of physicalism some kind of alien infection.

Ecofeminists gravitated to these sorts of tales of feminine harmony with nature—often alongside the gender essentialism and the nature-worshiping transphobia of figures like Mary Daly—yet many simultaneously embraced postmodernist themes and justifications. For example, Andrée Collard would write in *Rape Of The Wild* both that "*nature is a state of mind and cultural convention*" [28] and "*reality is the mind's eye.*" [29]

Antirealist philosophical currents in France and the US were, in this way, spreading and hybridizing with local incentives. As Nanda would go onto summarize in the case of India,

> "The twin themes of epistemic parity (modern science as only one among other ways of knowing, no better and often much worse), and populism (the right to live by one's own traditions), became the refrain of the critics of scientific temper... A whole generation of social activists and science studies scholars has emerged in India who take these two themes as axiomatic. Even those who have doubts in private, dare not challenge the consensus in public for the fear of being labeled as "positivists" and elitist... Inspired by Feyerabend, the self-proclaimed defenders of the common man argued that astrology was the myth of the weak, much as science was the myth of the strong."
>
> Meera Nanda, *Prophets Facing Backward*, 2004

In this sense, the importing of postmodernist denials of an objective reality was simply an opening maneuver to wipe away the dominant scientific models so that they could then be replaced by their own sweeping claims. Moon goddesses and period blood cults for the transphobic ecofeminists, astrology and quantum mysticism for the hindufascists.

In many ways, the embrace of postmodernist arguments by ecofeminists, postcolonial theorists, and finally reactionaries or outright fascists in India was a return home. Yes, *Tel Quel* had its origins as yet another maoist cult of white kids fetishizing poor foreign brown people, but also the popular anti-realist currents in the US and the western Left that empowered postmodernism as a movement had distant roots in currents of philosophical idealism within the vedic traditions now grouped together as "hindu." Specifically the Advaita Vedanta school, in which has long circulated the claim that the physical universe of matter is an illusion generated by underlying processes of spirit or mind. This imported kernel had been riffed on by European and American spiritualists and occultists for well over

---

[28] Andrée Collard and Joyce Contrucci, Rape Of The Wild: Man's Violence Against Animals and the Earth (1989), 5.

[29] ibid, 43.

a century, with great orientalist gusto.

To the western Left, the acclaim that postmodernist ideas received in places like India was intense validation of it.

As a consequence, the '80s and '90s were an era of epistemic devolution with new age spiritualism and conspiracy theories in wide circulation in the Left, Aronowitz, for example, thought nothing of confidently proclaiming on television that the light from computer screens causes cancer and scientific studies on such couldn't be trusted.[30] Such was not risky; it was a canned applause line.

Throughout the '80s and '90s, the anarchist mainstream remained committed to frameworks deeply at odds with the burgeoning cultic milieu and its antirealist legitimizers. One prominent anarchist journal at the time even titled itself *Reality Now!* As we will cover later, pretty much all the leading figures of the anarchist movement openly despised postmodernism, and those involved in daily projects or struggles simply ignored it. But its influence grew so vast in U.S. subcultural scenes as to become an unavoidable part of the Left.

As the child of an anarchist and a communist, I was dragged along to—and by my early teens autonomously went to—conferences, talks, protests, shows, org meetings, radical community spaces and lived with punks and activists across the radical Left during this period. From Indymedia nerds to forest defense hippies to squatting punks to solidarity union libertarian marxists. Especially as a teen, the flashy projects and actions of the anarchist youth groups I was a part of meant we were constantly being courted by adult activists of every stripe imaginable.

It was a deeply miserable time.

In the US, at least on the west coast, almost every Leftist was vehemently anti-vaccine, to the point where the topic was fearfully avoided by the few dissidents. Indeed, in Ellul's book where he argues for the conjunction of christianity and anarchism, one of his appeals to unity leverages the issue of vaccine rejection.[31] Hostility to fluoridation was mainstream. Identification with witchcraft, the occult, and various neo-paganisms was widespread to the point of almost full saturation. Plastic shamans like Lynn W. Andrews that had captured the '80s Left—hungry for the most grotesque noble savage narratives—were joined by the "Reclaiming Tradition," which played dress-up in the rags of a supposed European pagan history often invented from whole cloth. This was an era when the spiritualist Starhawk became widely cited and discussed as supposedly one of anarchism's leading theorists, despite doing shit like derailing the black bloc against the WTO in Cancun to corral people into a prayer circle instead.

By the time the Twin Towers fell, this sort of thinking was so deeply ingrained in the culture that half the west coast Left smoothly shifted to fervent 9/11

---

30 "Conversation with Stanley Aronowitz PhD," Conversations with Harold Hudson Channer, June 26th, 1995, uploaded February 11, 2020, Internet Archive, 46:45. https://archive.org/details/mnn_128_45781995.

31 Jacques Ellul, Anarchy and Christianity (1991).

trutherism. This all happened largely without critique, just as widespread misunderstandings of nuclear power and genetic modification were rarely openly critiqued because *"surely the state and Monsanto need no defenders!"*

Delighted misuse and misunderstanding of scientific theories to validate smug mysticism saturated virtually every conversation. The famous political prisoner Rob Los Ricos stood before a packed auditorium at Portland State University in the '00s giving a political presentation about how physicists lie and *"just make something up... Almost everything in the whole universe is plasma. It's not really matter, it's not really not-matter. People think the sun has all this matter and it's burning off energy. It's not!...The sun will never go out!"* [32] Lamborn Wilson excitedly proclaimed that quantum mechanics had endorsed, *"Anti-realism or even Berkelean Idealism."* [33] A friend and ally of his, the egoist writer Wolfi Landstreicher, in a screed against science, cited it as a supposedly known fact that quantum mechanics says our minds create the world through *"observing"* it.[34] By 1985 you can find it confidently declared in *Popular Reality* #7 (editor Bob Black was also close allies with Lamborn Wilson) and copied from the pages of Maximum Rocknroll that *"there is no 'real' world out there... Atoms don't exist until you look at them; reality is what we are taught it is."* [35] Todd May makes similar deeply misinformed appeals to quantum mechanics, and Godel's incompleteness theorem, by 1994 in *The Political Philosophy of Poststructuralist Anarchism*. Such mystical misunderstandings or misrepresentations of science were everywhere in the Left.

Postmodernism, the supposed *ideology of incredulity*, flourished amid this zeitgeist as the justifying philosophical insight that any deviation from epistemic diligence or a violation of the "establishment" perspective was good, but especially when it came to the physicalist model of reality given by science. The philosopher Giorgio Agamben's exhortation to his friend that he had a duty to reject the scourge of modern medicine and avoid a (life-saving) heart transplant was nothing notable at the time, it was the mainstream—almost mandatory—take.[36] At the core of the scene around *Semiotext(e)*, the prominent self-identified postmodernist author Kathy Acker rejected "mainstream medicine" to handle her cancer on ideological grounds, instead turning to psychic healers and herbalists;

---

32 "Law and Disorder @ PSU - Rob los Ricos Thaxton," Law and Disorder Conference at Portland State University, 2009, uploaded 2013, Internet Archive, 11:00 https://archive.org/details/bliptv-20131014-004423-Bmediacollective-LADRob-LosRicosH26424fps528.

33 Hakim Bey, Quantum Mechanics & Chaos Theory: Anarchist Meditations on N. Herbert's Quantum Reality: Beyond the New Physics.

34 Wolfi Landstreicher, "A Balanced Account of the World: A Critical Look at the Scientific World View," Killing King Abacus 2 (2001).

35 Jerod Poor, "Nihilism," Popular Reality 7 (1985): 3.

36 Adam Kotsko, "What Happened to Giorgio Agamben?," Slate, February 20, 2022, .https://slate.com/human-interest/2022/02/giorgio-agamben-covid-holocaust-comparison-right-wing-protest.html.

it did not go well.[37]

In some cases the appeals made were a stretch, like when Derrida's *"there is nothing outside the text"* would be—repeatedly, enthusiastically, and confidently—cited to me as an authoritative proof that there is nothing outside human sociality and thus no underlying physical world. But there were plenty of other arguments and texts that more directly reflected what the movement passionately wanted to hear. Indeed, the very term "postmodernism" itself was settled on by Lyotard only after passing over his first preference: *"paganism."*

---

[37] Kathy Acker, "The gift of disease," The Guardian, January 18, 1997, https://outwardfromnothingness.com/the-gift-of-disease-i-el-don-de-la-enfermedad-i/.

2

THE METANARRATIVE OF POSTMODERNISM

"Don't you see that criticizing is still knowing, knowing better? That the critical relation still falls within the sphere of knowledge, of 'realization' and thus of the assumption of power? Critique must be drifted out of."
Jean-François Lyotard, 1972, Driftworks

"Universally accepted notions such as unity, totality, and synthesis are re-examined and subsequently contested, thereby undermining not only science, but the rules of logic"
Marilyn August and Ann Liddle, (in praise for Bataille and Tel Quel) SubStance, 1973

"Secession, geopolitical disintegration, fragmentation, splitting—disagreement escapes dialectics and separates in space. Anti-universalism, concretely, is not a philosophical position but an effectively defensible assertion of diversity."
Nick Land, Against Universalism, 2016

It has become a popular strategy of retreat to collapse postmodernism down to one of Lyotard's quick descriptions as "skepticism of metanarratives," but of course "metanarrative" is an incredibly hazy concept. While it certainly makes some sense to speak of marxist accounts of history as a story of class struggle across progressing stages as a grand metanarrative, once you start viewing all models of reality within the amorphous metaphorical category of "stories" even things like the speed of light or Noether's Theorem quickly become "grand metanarratives" merely because they describes things and do so with "grand" universality. An anarchist can say "I think freedom is good and oppression is bad" and the postmodernist can smirk, "well I'm skeptical of that metanarrative." In this way the mere invocation of the term "metanarrative" shifts our conceptual and metaphorical registers to literature and all its associated linguistic games or frameworks, and what it ends up targeting for rejection is all generality or universality. Most people who haven't already been slowly acculturated within postmodernism are going to recoil from the "rejection of all descriptive generality or universality" when stated plainly, and so starting with the conceptual fuzziness of the literary frame is necessary to build towards this conclusion.

Now even this critique of the conceptual fuzziness of "metanarrative" is not going to land with most postmodernists because postmodernism is obviously not just "skepticism of metanarratives" and one of its most recurring premises is that concepts don't—and can't—have clean or tight definitions, nor ever cleave reality "at the joints," with many believing it's trivially preposterous to think our concepts can refer in any substantive sense to physical reality at all. If every concept is a hazy bundle of poetic associations, with precision and grasp upon reality as both impossible fantasies, then how dare I complain that "metanarrative" is a hazy bundle of associations.

This is part of the meta-justificatory framework that's often silently at play behind the slogan "skepticism of metanarratives."

It is a part of a grand historical narrative:

> Once there was this historical epoch called "modernity" where people credulously believed in metanarratives, but now we've progressed to a new stage of history wherein we've invented general skepticism of metanarratives. Once there was this mistaken enlightenment paradigm in Europe called "humanism" where people believed in things like "truth" and "freedom" and this directly

motivated all imperialism, racism, genocide and bad things in history, making it now the heroic and inevitable task of leftists to expunge all lingering traces of it. Once there was this silly thing called "positivism," which stands for anything positively inclined to science or truth, but it couldn't justify itself and is only retained by losers who haven't gotten the memo.

The triumphalism of the "post-" is a rhetorical maneuver that developed wide cachet in the mid twentieth century: rather than earnestly saying "you're wrong and we're right" the move is to instead sneer "we're part of the conversation that has *moved beyond*." Indeed in some wings of academia "post-" becomes synonymous with any contemporary understandings, and "modernism" becomes less a coherent concept to be rejected or embraced and more a garbage bin of a slur in which to stick every single outdated idea or insufficiently complex account. Indeed, to many younger generations of academics "postmodernism" is taken as a signifier of "complex" which in turn is the implicit goal and ideal of academic accounts (can't get a lot of essays out of a simple truth). In the absence of any defeater of a given paradigm, one simply runs up the flag of being "post-" that paradigm and then assigns prestige and social capital according to a person's investment in your new discursive community.

Ideologically speaking, this is a very marxist move in that it tries to bypass the explicit moral and metaethical conversation—inherent to other discursive spaces like anarchism—and instead leap to a historical account where there are determined winners and losers from the get-go and wouldn't you rather be on the winning team? There's also a slight variation where historical victory is less appealed to than *novelty*: the old discourses are played out, if you want to be interesting you have to keep up with the changing fashion.

Of course, this doesn't actually avoid engaging in normativity or metaethics—such is impossible, even an urge to climb the hierarchy of "coolness" is a value judgment—it just tries to obfuscate and stack the conversation so that assumptions in those domains avoid questioning. And there's a huge moral metanarrative also at play.

In first discussing the origins of postmodernism as a social movement, I couldn't help but mention the way that much of the Left latched upon an analysis of fascism as "totalitarianism" in the '50s and '60s. While the term "totalitarianism" originates with Mussolini's ghostwriter—the staunchly anti-realist philosopher Giovanni Gentile—it rose to prominence as a retroactive classification of fascism, primarily motivated by a need to critique the USSR and US by comparison. "Totalitarianism" then became an even more sweeping categorization to bring to the fore certain common convergences between these regimes: total social control under a vast total state with total social narratives with vast support. Certainly a bad thing! But in this lens, homogeneity and largeness thus became the primary characteristics associated with fascism, with its use of technology in particular marking the Third Reich/US/USSR as historically distinct from prior authoritarianisms and nationalisms.

Notorious "orgasm energy" wingnut Wilhelm Reich was profoundly influential in this era, and continued to be glowingly cited in journals from *Semiotext(e)* to *Social Text* and countless postmodernist zines into the '80s; and his account of fascism in terms of sexual repression and conformity resonated with many in the burgeoning sexual revolutions for obvious reasons. And in Hannah Arendt's even more influential account, totalitarianism is fundamentally a product of scale, a modern dynamic of vast superfluous masses, and thus she explicitly writes off the capacity for "small countries"[38] to become totalitarian and thus fascist. Authoritarian dictatorships of extreme state control, yes, but not true fascism. This would of course be a surprise to many early fascists and certainly most contemporary fascists who, across countless ideological currents since the '80s, have mostly converged on a demand for micro-nationalism, often down to the scale of villages or communes of less than 150 people. Just as it would be a surprise to Arendt's mentor and lover Heidegger—chief philosopher of the Nazi party—an ardent enemy of modernity and proponent of a return to the small scale, natural, and local.

Despite echoing many accurate critiques of the Nazi project that emphasized the importance of truth,[39] Arendt simultaneously perpetuated a framing of truth itself as totalitarian and "*despotic*"[40] because of the way that it *obliges* us, casting dishonesty as necessary for democracy (supposedly a good thing and somehow at odds with fascism). This account of "totalitarianism" as any absolutism beyond democratic negotiation and compromise quickly transformed into an attack on anarchism. Claude Lefort, a prominent French state-communist thinker who helped popularize the "totalitarian" critique, decried the classic anarchist aspiration of the "abolition of power" as "totalitarian" while also describing autonomy as "imaginary" and "*a realm governed by despotic thought.*"[41] In this account, the real slavery is consistently valuing freedom, because such universality is tantamount to fascism.

It should be obvious that this was a gravely mistaken project from the start. Attempting to define fascism in terms of its commonalities with the US and USSR (or more to the point, critiquing the USSR and US in terms of

---

38 Hannah Arendt, *The Origins of Totalitarianism* (1951), 310.
39 See her focus on fascism's erosion of *"distinction between fact and fiction… and the distinction between true and false"* in *The Origins of Totalitarianism*.
40 Hannah Arendt, *Between Past and Future* (1969), 241.
41 Claude Lefort, "Politics and Human Rights" in *The Political Forms of Modern Society: Bureaucracy, Democracy, Totalitarianism*, ed. by John B. Thompson (1986), 270.

their commonalities with a few regimes where fascists took power) inherently obscures what is distinct about fascism. It entirely ignores that ideologies aren't reducible to the policies or strategies that regimes under their banner adopted when in power (much less a mere two regimes within the same sliver of history). The nazis were fervent opponents of homogenization; they built their base on critiquing globalization and positioned themselves as the defenders of national difference and diversity, noble stewards of Europe out to help the little guys free themselves from the encroachment of global culture and universal ethical claims, to restore and preserve national uniquenesses. Fascism is nothing if not a militant declaration of *difference*.

That powerful states in the same period in history often converged upon similar strategies, technologies, or means—like industrialization, isolation, and mass politics—regardless of their ideologies or motivating ends, is a banality. So too does an implicit exceptionalism towards the Holocaust on the grounds that some aspects of it involved modern bureaucracy, miss deep commonalities with genocides throughout history, like those perpetrated by the Mongols. To define "fascism" only in terms of institutional or sociological dynamics in a single moment in time can hold one back from understanding how the underlying philosophy, value system, and movement will express itself differently in different historical conditions—whether within the halls of power or as insurgents without. It would miss, for example, the many branches of fascism today not seeking to seize state power but to topple technological civilization, launching a new age of micro-tribal warlordism and slaughter in the ruins.[42]

But this misanalysis of fascism had intense opportunistic rhetorical utility for many Leftists in the post-war period. The Left—in contrast to a specific ideology like anarchism—is not defined by any eternal core values or theoretical constructs, but is purely a sociological coalition of underdogs, as such its analysis at any given moment is shaped most strongly by the imperative to preserve and unify that coalition, avoiding anything thorny that might fracture it. And nothing preserves a mass coalition like a single monolithic and institutional enemy. In more stodgy and dated sections of the Left, this may look like a focus on unifying the working class against the capitalist class, dismissing things like racism or patriarchy except insofar as they are clear obstacles to working class cohesion or tools of capitalist rulership. To hold together such diverse mass coalitions necessarily involves the studious avoidance of coherence and consistency. In the pages of early issues of *Semiotext(e)*

---

42 See the influential fascist currents since the '80s of "leaderless resistance," "national anarchism," neoreactionary "patchwork," ecofascism, pan-secessionism, etc. with broader influence on anti-globalism, "multipolarity" etc.

this looked like denouncing the staunch anti-imperialist activist Bertrand Russell as actually an *imperialist* of reason, out to crush diversity.[43]

Totalitarianism presented a more useful polarity: when the ruling establishment was defined in terms of homogeneity, vastness, and generalness, the coalition of underdogs could be directly defined in terms of a common heterogeneity, smallness and particularity. This gained wide cachet in the Left because setting the unifying value as **particularity** promotes tolerance between underdogs. It is thus the core moral metanarrative that postmodernism depends on as a background assumption. That the ur-evil of fascism is a matter of generality or universality is so deeply rooted as to often go without conscious awareness. Some postmodernists may be happy to chuck the moral dichotomy of "freedom vs domination" as meaningless or poorly posed, but they will still violently recoil in defense of the alternative moral dichotomy of "particularity vs. universality." As Lyotard frankly declared in *What Is Postmodernism*: "*Let us wage war on totality.*" [44]

At the most brute level this is the reason folks like Peter Lamborn Wilson celebrated ethnic nationalisms—in the examples he gave in *Millenium*, "*Bosnian Muslim or Finnish or Celtic or Ashanti*"—as defending **difference**,[45] and it was how Foucault could embrace nationalisms in the global South, infamously praising the theocracy of Iran, an inane "anti-imperialist" campism in the same vein as the stalinists who later praised ISIS. In the topsy-turvy metanarrative created by the totalitarianism thesis, preserving the particularities of national character—no matter how reactionary—against global values is anti-fascist and those feminists complaining about being thrown on the bonfire are corrupted by the true fascism: western universalism. To complain about a local tyranny is to be a bad ally in the grand pluralist coalition of particularist underdogs.

In this way postmodernism welded the liberal value of **pluralism** to the campism of mass-coalition leftism. This basically allows for the reinvention of panarchy but devoid of any awareness of the longstanding anarchist critique of it, often in quite shocking ways. In *Poststructuralism and the Epistemic Basis of Anarchism*, Andrew Koch literally argues that this pluralism constitutes anarchism on the grounds that, "*the recognition of plurality becomes the basis for resistance to that which would impose universals,*" and thus the "legitimacy" of the state. Neglecting entirely, of course, that anarchism is not mere anti-statism, but a universal ethical imperative to attack power in all its forms at all scales,

---

43 Roger McKeon, "Gaiety, A Difficult Science,*"Semiotext(e)* III no. 1, (1997): 8.
44 Jean-François Lyotard, *The Postmodern Condition: A Report on Knowledge,* trans. Geoff Bennington and Brian Massumi (1979), 82.
45 Peter Lamborn Wilson, *Millennium* (1996).

including the interpersonal—as anarchists and antifascists firmly concluded in resisting the attempted entryism of "national anarchism" throughout the '00s: *a pluralistic patchwork of fascist tribes is trivially not anarchy*. But pluralism remains appealing for academics and leftists precisely because it can be wielded to patch over any contradiction, tension, or danger.

> "To pluralize is always to provide oneself with an emergency exit, up until the moment when it's the plural that kills you."
> Jacques Derrida, *Resistances of Psychoanalysis*, 1996

Because so many postmodernists lean so heavily on this kind of pluralism, it's common for folks to critique postmodernism as promoting cultural *relativism*, but this drastically understates the issue. Sure, postmodernists tend to assert cultural relativism in truth and ethics, but this is almost always motivated not by a neutral epistemic claim that two cultural epistemologies or moral foundations can differ so greatly as to be irreconcilable, but from a taken-for-granted background metanarrative that it would be a bad thing if cultures could reconcile towards universalism.

Heidegger infamously blamed the Jews for their own extermination in the Holocaust, claiming that the camps were the apex expression of universalism and technology (two things he despised as Jewish), and as such the Jews had justly reaped what they had sown.[46] Don't blame the anti-modernists that literally did it; in the grand scheme of things this was modernism eating itself! To their eternal embarrassment, much of this narrative continues to circulate in many humanities departments, where the Nazis are framed not as the apex exemplars of romantic irrationality but as the high water mark of the evils of reason, science, and "humanism."

This thesis is so wildly popular as to be a completely unquestioned background fact in many spaces. Dozens if not hundreds of folks have furiously told me far more extreme versions of the above over the last two decades along with the perennial "*So I hear you think there's such a thing as truth! Don't you know that whole notion is just western imperialism?!*" in flabbergasted sneers or earnest confusion from bookfairs to Food Not Bombs gatherings to radical dance parties. Always from people astonished that anyone could be so ignorant of the fact that thinking there's a single reality was directly and innately responsible for the invention of patriarchy and every genocide of indigenous people. And they're always the college sort, with all the wealth that connotes. A kid from poverty might in rare cases join in playing with witchcraft, chaos magick, woo, spiritualism, and astrology, but the truly ardent zealots of anti-universalism overwhelmingly share a relatively privileged class

---

46 Richard Wolin, *Heidegger In Ruins: Between Philosophy and Ideology* (2022), 78.

background.

To be fair, these conclusions usually aren't stated as extremely by tired adjunct professors as their students end up running with things in radical circles, but the gist is omnipresent, and said professors don't intervene and often approvingly nod along when their students leap ahead.

Anti-universalism as anti-totalitarianism as anti-fascism. Such conceptual conflation is the core normative thrust of postmodernism from which everything else grows.

In this frame, it makes sense to fetishize **transgression** as valuable in and of itself. If the foremost goal is to defeat the suffocating homogeneity of society, then edgelording is praxis. What's a little embrace of rape or pederasty? Why not invite a fascist on your podcast?! Why not compete over digging up the most shocking and scandalous texts for the reading group? Why not snarl at "cancel culture" by flagrantly violating whatever norm you can? Sure, I think that a given thing is bad, but that's besides the point, obviously the only thing that really matters is resisting the norms of society. The large scale is our enemy, and the small scale can never be an issue in comparison. This was particularly in vogue from postmodernism's gestation in France, during which era the Marquis de Sade was elevated from obscurity to a heroic national literary and philosophical figure, and it continues to the modern era where many postmodernism inclined folks in the scene I know still declare the band Death In June "inherently antifascist" because there's a gay member and everyone knows such transgression is the exact opposite of Real Fascism (which antifa represents, by picketing their shows). This current of cry-bully provocateurs are, as Habermas snarked, "*addicted to the fascination of that horror which accompanies the act of profaning, and is yet always in flight from the trivial results of profanation.*" [47]

It also follows naturally to worship **illegibility**. If the state requires some forms of legibility to rule, why that surely means legibility is the state in micro and illegibility is inherently anti-authoritarian and can never create or reinforce hierarchies by impeding accessibility. Mark Fisher once groused that in "poststructuralist continentalism" there's an assumption that "*the more laborious and agonised the writing, the more thought must be going on.*" [48] Certainly, when *Tel Quel* was accused of "illisibilité" they took it as a point of pride. And Foucault infamously broke ranks in sharing this opinion of the original French scene, in particular deriding Derrida as an obscurantist terrorist, "*He writes*

---

[47] Jürgen Habermas, "Modernity versus Postmodernity," *New German Critique*, no. 22, Special Issue on Modernism (Winter, 1981): 3-14.

[48] "Terminator vs Avatar," Mark Fisher in *#ACCELERATE: The Accelerationist Reader*, ed Robin MacKay and Armen Avanessian (2014), 341.

so obscurely you can't tell what he's saying. That's the obscurantist part. And when you criticize him he can always say, 'You didn't understand me; you're an idiot.' That's the terrorism part." [49] David Bloor likewise betrayed the tribe by harshly characterizing Latour's work as "*obscurantism raised to the level of a general methodological principle.*"[50]

Critics have laboriously teased out the mechanisms by which postmodernist obscurantism is encouraged and propagated,[51] but one doesn't have to appeal to the incentive of academic grifters chasing novelty and interpretive latitude, nor the followup pretentious gatekeepers falling into sunk cost fallacies and using inaccessibility to create status hierarchies; if you're trying to avoid universality it makes sense to deliberately avoid clarity of language as well as of thought. And make no mistake, quite explicit rejections of legibility are common in postmodernist circles, like in Denise Ferreira da Silva's repeated hostile references to "transparency." Mantras I heard throughout the '00s Left like "to define is to limit" and "definitions are fascist" might seem like wild generalizations of state control patterns in limited contexts, but they make sense as applications of the grand metanarrative where clarity is dangerously close to universality, and thus fascism. Nevermind that such arguments were then aggressively leveraged against the exclusion of Troy Southgate's neonazi "national-anarchists" from our spaces. To quote what Aragorn Moser—a former maoist and disciple of Marcuse—once said, in person, about refusing to kick a neonazi "Bay Area National Anarchist" member from anarchist spaces, "*Nothing could be more authoritarian than to define anarchism.*"

> "Holding a position with conviction is seen as unpleasantly authoritarian, whereas to be fuzzy, skeptical and ambiguous is somehow democratic."
> Terry Eagleton, *After Theory*, 2003

With this metanarrative of resisting the imperialism of universality comes an intense valuation of **humility**. This is shared with relativists across a much wider expanse of liberal academia where social norms strongly pressure everyone to, as Martin Kusch put it, "*oppose intellectual imperialism and value epistemic plurality and tolerance.*" [52] Pluralism is about suppressing conflict between coalition allies and one of the best ways to do that is to aggressively

---

49 Steven Postrel and Edward Feser. "Reality Principles: An Interview with John R. Searle," *Reason Magazine*. February 1, 2000. https://reason.com/2000/02/01/reality-principles-an-intervie/

50 David Bloor, "Anti-Latour", *Studies in History and Philosophy of Science Part A*. 30, no. 1 (March 1999): 81–112.

51 Filip Buekens and Maarten Boudry, "Psychoanalytic Facts as Unintended Institutional Facts," *Philosophy of the Social Sciences*. 24, no. 2 (2011).

52 Martin Kusch, *Relativism in the Philosophy Of Science* (2020).

police anything that looks like uppity claims extending beyond one's limited domain. Wherever you have a fixed upper class, they tend to quickly recognize the necessity of performative humility (as well as interpretative latitude in statements). The nouveau riche might be crass and arrogant, and the poor—usually more tightly tied to objective materiality—might be obliged to speak plainly and directly of practical factuality, but today's ossified and comfortable middle classes as well as many prior aristocratic classes, prefer indirectness. Exactly when the bus comes and the most efficient way to repair your car's engine are facts of too much importance to duck and weave around; an argument about such should be clear and forceful. But when what's on the line isn't survival but status, the dominant strategy that emerges is performative humility (and aggressive censure of anyone who deviates). "You're objectively fucking wrong," is praiseworthily honest behavior when arguing about how to repair the restaurant's dishwasher (or solve a path integral), but scandalous aggression in an art gallery or most humanities departments.

Often, but not always, implicit in all this is a framework of **negative liberty**, that is to say *freedom from* as opposed to the positive notion of *freedom to*.

Freedom from is about preserving separate distinct patches, where everyone can be king (or expert) of their limited domain. Whereas *freedom to* is an arrow pointing dangerously outward to infinity and thus the universal. *Freedom from* often assumes a zero-sum world, whereas the framework of *freedom to* embraces positive-sum collaboration, where our freedoms can compound and expand through interaction rather than terminating, as the saying goes, where your fist ends and my face begins. Amusingly—given the lofty language often used in the more transgressive veins—the more negative orientation here of postmodernists can lead to a rejection of desire. So like in Saul Newman's case, the postmodernist metanarrative ironically leads to a certain degrowthy stoicism that rejects desire because desiring is seen as a product of capitalism and necessary to it.[53] In discussing the intentionality of various sorts of claims to knowledge, Aronowitz moralizes about a distinction between those encouraging passive "*acceptance*" (good) and those that enable more efficaciously moving in or changing the world (bad). In Duane Rousselle's case, having infinite or limitless desire is literally characterized as "*tyranny*."[54] Since freedom is implicitly seen as avoiding external influences (e.g. avoiding being shaped rather than avoiding being chained) and oppression is seen as intruding beyond your local patch, it follows that one should quiet one's hungers, abandon the infinite (like Malatesta's neverending hike towards anarchy),

---

53 Saul Newman, *Postanarchism* (2016), 59.
54 Duane Rosselle, "Post-Anarchism and Psychoanalysis", *Cyber Dandy*, uploaded July 28, 2023, YouTube, 29:10, https://www.youtube.com/watch?v=S0pz5SBHIQ8.

and learn to be content at home with what one has. Rather than the standard anarchist cries of supercharged desire—"*not just bread but the bakery and the entire heavens!*" and "*we want everything!*"—the prescription is instead a kind of retreat away from the world. In a related dynamic, many leftists and radicals have long critiqued currents in postmodernism for devolving into political quietism and retreat from struggle, a criticism that was emphasized by many radical leftists in Sokal's circles.

Now it goes without saying that people derive varied prescriptions and focuses from this metanarrative of opposition to universalism. Someone fetishizing humility might see no appeal to transgression, and vice versa. There are a few writers who are starkly earnest and far from obscurantism, like Jesse Cohen. Moreover, the metanarrative can partially break when crashing against existing ideals. Many feminist and anarchist theorists mixed up in the postmodernist wave immediately realized that there was no way to duck the universalist charges and remain feminists or anarchists in any remote sense, sometimes quite frankly admitting in their texts the necessity of a universal moral value or a single objective material reality, explicitly contra many of their postmodern allies. For example Sarah Ahmed[55] or Todd May, who, in *The Political Philosophy of Poststructuralist Anarchism*, tries to insert an analytic account of core universal ethical principles into Foucault, Lyotard, and Deleuze to acknowledge that they were moralizing the whole time and that any attempt to advocate an escape from morality is incoherent as well as directly adversarial to anarchism. Most will take arguments or values pretty far, but still recoil from raw conscience when their ideological frame would otherwise lead to supporting a fascist. It's common across any ideological movement to see patchwork ideological or philosophical positions, that accept some conclusions and reject others, while keeping the broad metanarrative afloat. Even within the original French scene, Deleuze was a rebel who stanned for alien philosophers in the "analytic" vein like Hume and Whitehead.

But there are still some quite clear commonalities. And if the triumphalist narrative of "once there was universalism, which unleashed the evil of fascism (largeness and sameness), and now we have moved beyond such mistakes" could license the fetishization of humility, illegibility, transgression, and retreat, its most notable connection was to **antirealism**, the contention that launched the The Science Wars.

Before we dive into exploring antirealism in all its philosophical forms, let's return to the historical narrative we left off.

---

55 Sarah Ahmed, *Differences that Matter: Feminist Theory and Postmodernism* (1998).

3

THE WAR, THE HOAX, AND THE NARRATIVE

"A friend who teaches an Introduction to Pomo course told me that even after hearing that Sokal's article was a parody and then re-reading the piece, she nonetheless still thought that the article was excellent and instructive."
Michael Albert, *Sokal 1*, 1996

"To put it succinctly, had I been shown both Sokal's and Aronowitz's articles, and asked which might be a hoax, I would have said that both are either hoaxes or nonsense! Why? Because Aronowitz, like Sokal, (1) makes statements about physics that are factually wrong, (2) displays deep misconceptions about physics and (3) seems ignorant about what physicists did in the past and try to do now. And to boot, Aronowitz, like Sokal, speaks with a self-confidence that would assure lay readers that they are in the hands of an erudite expert."
Kurt Gottfried, *Was Sokal's Hoax Justified?*, 1997

"One friend of mine told me that Sokal's article came up in a meeting of a left reading group that he belongs to. The discussion became polarized between impassioned supporters and equally impassioned opponents of Sokal; it nearly turned into a shouting match. The astonishing thing about this, my friend said, was that actually no one had read the article, because that issue of Social Text had sold out so quickly. Members of this group knew about the article only from having read accounts of it in the press, or from discussions with others who had read it."
Barbara Epstein, *Postmodernism and the Left*, 1997

As postmodernism and associated antirealist philosophies rose in number and influence in both the radical left and academia, so did aghast reactions. Physicists were notably riled, with, for example, Theo Theocharis and Michalis Psimopoulos aggressively critiquing the *"betrayers of the truth"* by 1987.[56] But philosophers were also strongly critical, and, by 1990, when Larry Laudan published his influential book *Science And Relativism* summarizing the conflict, it had been going for a while.

In 1994, the biologist Paul R. Gross and the mathematician Norman Levitt published *Higher Superstition* in which they complained at length about postmodernism and some related currents popular in left-wing academia. Much of the book is kvetching about the supposed abandonment of enlightenment universalism for relativism and outright denial of science, complaining that people like relativist sociologists and Aronowitz, *"reduce the scientific enterprise to little more than culturally-determined guess work at best and hegemonic power mongering at worst."*[57] But the book also, at points, pandered to right wing complaints about identity politics, and—perhaps as a result—sold quite well, finding populist success for a complaint about academia.

Higher Superstition is now often talked about as the genesis of the Science Wars, but widespread opprobrium strongly predated it. Months before its publication, the then quite prominent anarchist Murray Bookchin excoriated postmodernism at length, complained of the collapse of realism into mysticism, and specifically singled out Feyerabend's attacks on science:

> "In recent decades, both in the United States and abroad, the academy and a subculture of self-styled postmodernist intellectuals have nourished an entirely new ensemble of cultural conventions that stem from a corrosive social, political, and moral relativism. This ensemble encompasses a crude nominalism, pluralism, and skepticism, an extreme subjectivism, and even outright nihilism and antihumanism in various combinations and permutations, sometimes of a thoroughly misanthropic nature....Paul Feyerabend's corrosive (in my view, cynical) relativism to the contrary notwithstanding, the natural sciences in the past three centuries have been among the most emancipatory human endeavors in the history of ideas — partly because of their pursuit of unifying or foundational explanations of reality. In the end, what

---

56 Theo Theocharis and Michalis Psimopoulos, "Where science has gone wrong," *Nature* 329 (1987): 595–598.
57 Michael Shermer, "Farewell to Norman Jay Levitt (1943–2009)," *eSkeptic*, October 27, 2009, https://archive.skeptic.com/archive/eskeptic/09-10-26/.

should always be of concern to us is the content of objective principles, be they in science, social theory, or ethics, not a flippant condemnation of their claims to coherence and objectivity per se."

Murray Bookchin, *The Philosophy of Social Ecology*, 1996

At this moment Bookchin was a central figure in a vast and acrimonious debate across anarchism and the radical Left, known as a fight between "organizationalists" and "lifestylists," social ecologists vs. deep ecologists, or just "reds" and "greens," depending on one's ideological lens or charitability. Bookchin was notoriously crude in lumping together a variety of anti-modern subcultural strands, but his recognition of strong overlap between postmodernism and new age mysticism was hardly novel. Many were quite openly emphasizing the overlap as positive, like Carmen Kuhling in *The New Age Ethic and the Spirit of Postmodernity*.

Yet in one important respect, Bookchin was utterly and pretty trivially wrong. Despite his attempts to besmirch his arch-enemy—the only other anarchist alive at the time who rivaled him in fame and influence, the primitivist John Zerzan—he was *also* a fervent critic of postmodernism, and likewise on the grounds that it was undermining a necessary realism about the physical universe.

Indeed, although Bookchin quite evidently hadn't read it, Zerzan had written *The Catastrophe of Postmodernism* in 1991, critiquing the linguistic turn generally, scoffing at the relativism of figures Foucault and Lyotard, decrying the erosion of truth and reason, attacking postmodernism's *"thoroughly relativized academic sterility,"* and lamenting that in postmodernism, *"nature has been so far left behind that culture determines materiality."*

Despite Bookchin's attempt to slur the green anarchists he was in conflict with, very few in the anarchist milieu beyond the immediate circles of *Semiotext(e)* had anything nice to say about postmodernism.[58]

Indeed, while the anarchist movement fought vicious internal battles throughout the '90s, the one thing all sides kept interrupting their hostilities to emphasize was their hatred for the postmodernists and the denial of physical reality or all access to it.

> "Postmodernism... is exactly what Heidegger would have had in mind if he had stuck around long enough to see it. I think that here we have a rather complete abdication of reason with postmodernism in so many ways... They feel that the idea of origins is a false one (these are all big generalizations; there are prob-

---

58 For a sampling of some particularly spicy hostility to the postmodernists see Randall Amster, *Anarchism as Moral Theory: Praxis, Property, and the Postmodern* (1998). Sharif Gemie, *Habermas and Anarchism* (1991), And much of Miguel Amorós but particularly, his latter work, *The Golden Mediocrity* (2015).

> ably some with slightly different emphases). We are in culture. We've always been in culture. We always will be in culture. So we can't see outside of culture."
> John Zerzan, *Against Technology*, April 23, 1997

> "Latour makes a quantum leap to assert that scientists do not live up to their claims of practicing an orderly methodology. Facts are 'socially constructed', not discovered, he concludes, as a consequence of the microrelational give-and-take that makes up laboratory routines; their validity seems to hinge more on subjective interplay in the social world than the realities (if any) of the natural world. This conclusion seems to support the postmodernist notion that reality is actually chaotic and is only organized by disorderly scientists into orderly schemes... Stripped of their postmodernist verbiage, Latour and Woolgar almost pride themselves in acknowledging that their work is merely a fiction. Inasmuch as they offer no criteria at all by which to judge our suppositions about the natural world, we are deprived of all 'preexisting order' as a basis for formulating truthful statements about reality."
> Murray Bookchin, *Re-enchanting Humanity*, 1995

Anarchists were hardly alone in this. Across the radical left—which had firsthand experience with the postmodernists during their rise in the '80s—the resulting hatred was strong. Marxists wrote endless books and screeds, and so too did leftists like Barbara Ehrenreich, who would later shoot to fame with *Nickel And Dimed*, but was then an adversary of Aronowitz' in the DSA. Infamously, the leftist director Errol Morris quit his graduate program in rage at his teacher Kuhn's relativism and perceived antirealism.

This compounding hostility to postmodernism and broader antirealism on questions of physicalism is no small part of the context in which, in 1996, *Social Text* published a feature issue intended to circle the wagons.

Co-editor Andrew Ross introduced the issue combatively and set the overall tone of the counter-narrative he was trying to pitch in collaboration with Aronowitz: The storm of criticism they faced was really just a continuation of the same conservative attacks on other leftist academics that had broken out in the early '80s over the rise of minority studies and more diverse representation in books. An attack on postmodernists should thus be recognized by every leftist in academia as an attack upon *them*.

The upset leftist scientists were just hoodwinked fools, looking for someone to blame for their big particle collider project getting defunded. Aronowitz even declared that the scientists criticising postmodernism were no doubt actually motivated by deep fear of AIDS activists. Both were generally at pains to avoid conceding that any of their adversaries were leftists or radicals, but in Ross' one admission of this he quickly implied they were surely being warped by conservative funding.

In Ross' reframing, the outrage wasn't really about any philosophical differences over antirealism or irrationalism, it was about little humanities professors daring to assert the values of *democracy* against the authoritarian technocratic expertise of scientists. In the most audacious summary, he declares that those being criticized by scientists as irrationalists and antirealists want merely to make science more objectively accurate, to restore its ideals, to be self-critical, and to give the marginalized and affected more *democratic* say in the processes of science. This refusal to acknowledge the existence of the antirealists being critiqued or their positions and to instead slide in the most innocuous stuff that no one really objected to, was a potentially devastating rhetorical move. The signifier of "democracy," in particular, held significant currency in the era immediately following Students for a Democratic Society as many in the New Left settled into academia (remember that Aronowitz was a signatory of SDS' Port Huron Statement). It was enough of a cipher that anyone could read anything they wanted into it. Something generically positive, a common flag for progressives and the Left. Surely we would all agree about supporting democracy.[59]

Yet for all of the bluster about democracy being the core issue, Ross did respond to the charge of relativism… by arguing that epistemic relativism was not enough and that instead a focus on a more proactive embrace of the value of *diversity* was needed that would actively challenge the hegemonic footprint of science:

> "Nor can we be satisfied that the 'successes' of the lab scientist, the Chinese doctor, and the rainforest shaman are equally relevant and adequate to the cultures they serve. The power and authority of the Western scientific method alone have a global reach… Again, the relativist can show us how Western technoscience becomes acculturated and syncretized in other parts of the world but cannot insist that alternatives to this one-way process should be encouraged. Once it is acknowledged that the West does not have a monopoly on all the good scientific ideas in the world, or that reason, divorced from value, is not everywhere and always a productive human principle, then we should expect to see some self-modification of the universalist claims maintained on behalf of empirical rationality. Only then can we begin to talk about different ways of doing science, ways that downgrade methodology, experiment, and manufacturing in favor of local environments, cultural values, and principles of social justice. This is the way that leads from relativism to diversity."

---

59 For a good representation of the plumbline anarchist critique of Democracy see Crimethinc's *From Democracy To Freedom: The difference between government and self-determination* (2017). For a survey of anarchist critiques of democracy see *Anarchists Against Democracy: In Their Own Words*. For my lengthy critique of every different definition of democracy see William Gillis, "The Abolition Of Rulership Or The Rule Of All Over All?" *Human Iterations*, June 12, 2017, https://humaniterations.net/2017/06/12/the-abolition-of-rulership-or-the-rule-of-all-over-all/.

You might get the impression from this sneering that the physicist critics Ross was concerned with, like Gerald Holton, actually believed something like the claim that "the West" has a monopoly on good scientific ideas!

Beyond the talk of subjugating reason beneath some other values (in congruence with his talk about subjugating science under democracy) Ross quickly turned to bundlings of standard bearer political issues like *local manufacturing* with *local methodology* and *local values* or even *local principles*. In short, not a sharp or direct refutation of the relativist charges, but an attempt to defensively grab onto any tenuously relatable political associations.

Still, Ross was clear to grouse that the science warriors misrepresented their targets as:

> "boffo nihilists who deny outright the existence of natural phenomena such as recessive genes, or subatomic particles, or even the law of gravity."

Yet while Ross and Aronowitz' hostile screeds were meant to frame the feature issue, their attempts to draw in wide circles of support across in academia left them with contributions that were by comparison tame or unaligned.

Many scholars who answered their call took far more conciliatory approaches towards the scientists's criticisms, directly validating their concerns about issues of realism, and the biologist Richard Levins made a point to endorse scientific realism, saying that science, *"really does enlighten us about our interactions with the rest of the world, producing understanding and guiding our actions"* while defending social constructivism in its most trivial and unobjectionable form.

This was not exactly the kind of pugilistic support Ross and Aronowitz had hoped to rally, nor did it perfectly hew to their overall narrative about fighting for democracy against conservative scientists. But it was clean. No one denied that gravity represented a real material regularity deeper than any social convention.

If the charge of Gross and Levitt was that the *entirety* or even most of the academic left had been taken over by postmodernists, the scope of these contributors to *Social Text*'s feature issue was sufficient to disprove at least that. One could even walk away under the impression that no one challenged physical or scientific realism, that no one in left academia would even platform such absurdities, and that the storm of outrage must surely just be grounded in questions of democracy versus a would-be expert class.

But *Social Text* couldn't help themselves.

There was a particularly combative submission they'd gotten that really did come out swinging. The leftist physicist Alan Sokal, who had taught for the Sandinistas, had sent them a piece titled, "Transgressing the Boundaries:

Towards a Transformative Hermeneutics of Quantum Gravity."
This was Sokal's opening paragraph:

> "There are many natural scientists, and especially physicists, who continue to reject the notion that the disciplines concerned with social and cultural criticism can have anything to contribute, except perhaps peripherally, to their research. Still less are they receptive to the idea that the very foundations of their worldview must be revised or rebuilt in the light of such criticism. Rather, they cling to the dogma imposed by the long post-Enlightenment hegemony over the Western intellectual outlook, which can be summarized briefly as follows: that there exists an independent world, whose properties are independent of any individual human being and indeed of humanity as a whole; that these properties are encoded in eternal physical laws; and that human beings can obtain reliable, albeit imperfect and tentative, knowledge of these laws by hewing to the objective procedures and epistemological strictures prescribed by the (so-called) scientific method."

What followed was a long denunciation of this "dogma" via plainly silly arguments peppered with copious quotations from postmodernism aligned academics, and in particular editor Stanley Aronowitz. Indeed, so copious were the citations to Aronowitz that the article in many ways functions as a fisking of all of his worst claims. While the article rambled in focus, leaping across different subjects in a vapid but authoritative style mocking Aronowitz' own—complete with hilariously incorrect summaries of other thinkers—and clearly had fun injecting technical gibberish and new age concepts like "*the morphogenetic field*", the overall thrust of the piece could not have been clearer: Sokal declared quite frankly that physical reality itself, not merely our theories of it, "*is at bottom a social and linguistic construct.*"

While the physics references ensured that any physicist would immediately understand he was not serious—ensuring he couldn't later be accused of actually believing what he had submitted and then walking it back—Sokal hoped the vapid argument structures, ideological fawning, and appeals to authority, would be sufficiently apparent to the editors that they would reject it.

They chose to platform it.

The demonstration was devastating. *Social Text* could scoffingly imply that they were not denying the existence of an independent physical reality or in league with anyone doing such, but they were happy enough to platform an editorial very explicitly declaring exactly that.

> "Repeatedly throughout the editorial process, I asked the reviewers for comments, criticisms, and suggestions; I got none... I also employed some other strategies that are well-established (albeit sometimes inadvertently) in the genre: appeals to authority in lieu of logic; speculative theories passed off as established science; strained and even absurd analogies; rhetoric that sounds

good but whose meaning is ambiguous; and confusion between the technical and everyday senses of English words."

Alan Sokal, (in email to Michael Albert) *Sokal 1*, 1996

If *Social Text* had hoped to avoid any direct confrontation over realism, redirecting to a contrived dispute over democracy instead, that strategy had failed.

The wars that followed are, of course, common knowledge.

We all remember the horrific scenes of the Science Wars broadcast live by CNN embedded journalists during the '90s. With Aronowitz martyred, the SSK forces launched a series of bombings that were hesitantly backed up by broader STS allies. With the SSC forces losing their government sponsors, many physicists had no choice but to rob banks and go underground. Over a thousand assassinations, arsons, and targeted killings soon followed, and entire regions of academia made declarations of independence and alliances, creating complicated fracture zones. The cleansings and mass-relocations led to vast refugee camps in other departments, where food, sanitation, and grant-assistance was minimal. NATO airstrikes were much covered but did little to bring the situation under control, although for a while they stabilized alumni donation rates. Local warlords and provosts quickly began operating with only nominal connection to ostensible ideology. Reports widely circulated of outside sponsorship and infiltration, with heavy orchestration and involvement by the French Foreign Legion. As the conflict ground into a multi-year slog, factions pivoted in alliances in a complicated dance that infamously led to Liberation Against The Old Universalist Reactionaries (LATOUR) declaring that they had switched sides entirely. By the early '00s, peace and reconciliation committees were attempted, and special autonomous status was granted to certain surviving enclaves. Over a million people had perished, with reciprocal mass-killings on both sides, and entire generations were scarred with trauma. But a report by the UN Special Rapporteur cleared all parties of the highest war crimes alleged: miraculously not even a single professor had lost tenure.

Of course, in reality, what happened was Sokal's prank made the front page of newspapers across the world, everyone laughed at *Social Text*, and the postmodernists scrambled. Soon after Sokal published the book *Fashionable Nonsense* with the French physicist Jean Bricmont, listing examples of misused scientific ideas conjoined with garbled obscurantist prose, and then *Beyond The Hoax* by himself, documenting examples of antirealism and making philosophical critiques of them.

His targets seethed. Aronowitz, always happy to turn pomposity up to

11, declared that Sokal was simply "*half-educated*."[60] Julia Kristeva, a former maoist who had been a large part of *Tel Quel*, furiously decried Sokal and Bricmont, saying they should undergo psychiatric treatment.[61] (So much for postmodernist opposition to psychiatry!) Derrida referred to the critics as "*censors*."[62] And Lotringer, for his part, spat that Sokal was a "*cop*" and a "*dinosaur with a very small head*".[63]

This exercise in sneering helped the postmodernists affirm their collective loyalty, but it was highly ineffective beyond their ranks.

> "Let's concede once and for all that we are arrogant, mediocre, sexually frustrated scientists, ignorant in philosophy and enslaved by a scientific ideology (neoconservative or hard-line Marxist, take your pick). But please tell us what this implies concerning the validity or invalidity of our arguments."
>
> Alan Sokal and Jean Bricmont, *Fashionable Nonsense* (2nd ed.), 1999

And not everyone stuck to the party line of derision. Bruce Robbins, another editor at *Social Text*, who had been less pugilistic than Ross or Aronowitz, and by his account, got into a harsh six hour argument with the entire rest of the collective, chose to be more honest and self-reflective about why they'd published Sokal's piece, frankly admitting that it did do what Sokal claimed it proved, specifically, "*we thought it argued that quantum physics, properly understood, dovetails with postmodern philosophy*"[64] and "*enthusiasm for a supposed political ally*" in the fight had produced a blindness to the deficiencies of the content.[65] (Robbins would later leave *Social Text* and even collaborate closely with Sokal writing a prominent letter together as fellow Jews hostile to US support for Israel.[66]) But in the immediate aftermath, Robbins continued to insist on the political stakes of those philosophical questions of antirealism.

---

60 Editorial, "Professor Sokal's Transgression," *The New Criterion*, June, 1996, https://newcriterion.com/article/professor-sokals-transgression/.

61 Ulderico Munzi, "'Francesi, intellettuali impostori': Americani all'attacco di Parigi," *Corriere della Sera*, September 26, 1997, http://www.symbolic.parma.it/bertolin/ms11.htm.

62 Jacques Derrida, "Sokal et Bricmont ne sont pas sérieux" *Le Monde*, November 20, 1997, https://web.archive.org/web/20090216071032/http://peccatte.karefil.com/SBPresse/LeMonde201197Derrida.html.

63 Kristina Zarlengo, "Idiot Savants?" *Salon*, November 2, 1998, https://www.salon.com/1998/11/02/cov_02feature/.

64 Jay Rosen, "Swallow Hard: What Social Text Should Have Done," *Tikkun* (September/October 1996): 59-61.

65 Bruce Robbins,"Anatomy of a Hoax," *Tikkun* (September/October 1996): 58-59.

66 Jeffrey J. Williams, "Actually Existing Cosmopolitanism," *LA Review of Books*, December 23, 2018, https://lareviewofbooks.org/article/actually-existing-cosmopolitanism/.

> "Poststructuralism and postmodernism emerged at roughly the same time (the 1960s and 1970s) as the women's and civil rights movements at home and movements of national liberation around the world. Many would argue that their conceptual challenge to notions such as objective truth reflects and extends the political challenge of those movements."
>
> Bruce Robbins, Anatomy of a Hoax, 1996 [67]

And another *Social Text* editor found it hard to believe that Sokal *couldn't* believe in the antirealist claims in his article—such that the editor openly "*suspected that Sokal's parody was nothing of the sort, and that his admission represented a change of heart, or a folding of his intellectual resolve.*" [68]

Yet directly addressing the extent of these philosophical flights and the depth of their association and complicity was generally not in the interests of *Social Text*'s circle. The more the public heard about gravity being no more firm or grounded in pre-social physical reality than the rules of baseball, as the Executive Director of Duke University Press and publisher of *Social Text*, Stanley Fish put it,[69] the worse their image got.

Instead, under the crucible of ongoing worldwide mockery, these early approaches were discarded and a new collective strategy slowly congealed: Saturate every exchange with the reiteration of basic leftist points, thus implying your adversary disagrees and obliging them to waste time pointing out they agree while leaving them little to box with. Present the philosophical issues as boring and arcane minutia of no importance, nothing new and nothing affecting the real world. Avoid any reference to the starkest postmodernist antirealists. Any direct arguments for antirealism would be left to the STS scholars, who were closer to the physicists in culture and would defend more limited and nuanced claims in abstruse academic venues.

Meanwhile, *Social Text* worked to aggressively remove Sokal's article from circulation, excluding it from collections but refusing to allow anyone to reprint it, declaring they held the "copyright" alone. This censorship and brazen leveraging of the state's violence had the effect of stripping out the particularity of Sokal's text, and in particular the copious damning citations of the *Social Text* postmodernists and the stark declarations of antirealism. By making everything about "the hoax" in the abstract and suppressing popular circulation of the main text, the postmodernists thus managed to wrest some control back over the popular narrative. This meant deceptive slights of hand to frame the hoax as having been about merely sneaking erroneous technical

---

67 Bruce Robbins,"Anatomy of a Hoax," Tikkun (September/October 1996): 58-59.
68 Andrew Ross and Bruce Robins, "Mystery Science Theatre," *Lingua Franca* (July/August 1996): 54-57.
69 Stanley Fish, "Professor Sokal's bad joke," *The New York Times*, May 21, 1996 https://www.nytimes.com/1996/05/21/opinion/professor-sokal-s-bad-joke.html.

details (or "errata") past too naively trusting editors in a different field. *Social Text* could beat the heat for endorsing and platforming batshit antirealisms in their feature issue on the Science Wars and instead make everything about the humorously misrepresented science and how the poor literary professors took the deceptive physicist at his word.

Eventually Ross' original counter-narrative, that this was not an attack on the crimes of postmodernism specifically but upon *all* leftist academics, was significantly bolstered as right-wing media continued to harp on the story. Both *Social Text* and conservatives had a shared interest in equating postmodernism with the entire academic left. Anarchists and radical currents outside of academia were erased from all reference, and within academia the simmering tensions of the reversed two cultures polarizations were leveraged to increasingly imply that it was on some level *impossible* to be a scientist and a leftist or radical. To be a realist physicist was thus not just coalitionally suspect but somehow *definitionally* in conflict with egalitarianism. An opponent to postmodernism could at best be, as it started to be termed, a "left conservative."

As leftists in the humanities united under this metanarrative, the existential terror of mass critique from without started to diminish and the postmodernists were able to return to a more comfortably smug footing. All this pushback to postmodernism was the work of conservatives, and all they'd proved in this uncivil and tawdry stunt beneath the dignity of any true academic was a banality everyone already knew: some journals had low standards.

This narrative was helped along by Sokal's decision to limit his focus in his followup book, *Fashionable Nonsense,* to cataloguing arguable misapplications of scientific concepts, saving direct philosophical attack on antirealist claims for a later book, *Beyond The Hoax.* Sokal felt he was on firmer ground first establishing the widespread misapplication and misunderstanding of scientific theories in a subsection of the humanities, but this proved a comedic failure in many respects. Collecting instances of misused scientific terminology doesn't amount to much when the retreat is always available that such was just a poetic gesture. What gives physicists monopoly over the associations of a term like "*momentum*" anyway? Moreover, such drilling into terminology inherently came across as finicky and pedantic, the sort of petty point-scoring by those who refuse to engage with broader texts and discursive context. To give such sophomoric "critiques" any validation would threaten the entire project of the humanities.

If only Sokal had actually read some literature or engaged with the ideas of another field rather than just barging in and only proving that literature professors don't know the latest advanced physics. Oh well. At this point the thermonuclear Derrida quote is deployed.

> "This is all rather sad, don't you think? For poor Sokal, to begin with. His name remains linked to a hoax—'the Sokal hoax,' as they say in the United States—and not to scientific work."
>
> Jacques Derrida, comments to *Le Monde*, 1997 [70]

To this day, this remains the modern summary across many of the academic circles affected, as taught to their students and they to their students. As distance from any particulars or even awareness of the actual philosophical content has grown so has a tendency to re-frame the event as not just an attack on the Left as a whole but as a test of *peer review* (despite Sokal repeatedly saying the hoax wasn't about peer review, that he liked most of the content normally published in *Social Text* and approved of it not operating on peer review). It's just jaw-dropping.

> "In one case (the 'Sokal affair'), some scientists became famous for attacking gender and other areas of cultural studies as disciplines by sending a faked paper that trafficked in language developed in feminist theory to an academic cultural studies journal in the hopes of proving that a respectable publication would let even garbled work past peer review."
>
> Chanda Prescod-Weinstein, *The Disordered Cosmos*, 2021

Everyone knows peer review in modern academia is a faulty system with bad incentives, where tons of irrelevant journals basically auto-publish. Thus the narrative that the Sokal hoax was about exposing weak peer review allows one to point to examples of fraudulent claims in various scientific fields *as basically the same thing*. One heavily viewed YouTube video by a popularizer of STS even used the line of attack that Sokal didn't do a rigorous experiment with multiple submissions across multiple journals.[71]

I want to pause for a second here and point out how stupendously *weird* this defense is. When a leninist blog is caught platforming a submission about how "the Jews" created capitalism and the Holocaust was faked no one would ever object that *actually* this needed to be run multiple times with control groups or nothing can be validly extrapolated. The platforming of such foul reactionary ideas is in-and-of-itself instantaneously discrediting. If the blog had previously dabbled in a variety of borderline antisemitic tropes or been critiqued for repeatedly using "zionist" in ways not targeted at Israeli ethnonationalist settler colonialism but functionally implying Jews as a whole, we would rightly draw a connection. If the piece had been submitted on false pretenses by an antifascist group doing an infiltration operation, we would

---

70 Jacques Derrida, *Paper Machines*, trans. Rachel Bowlby, (2005), 70.
71 *"the physicist who tried to debunk postmodernism,"* Dr. Fatima, uploaded April 11, 2024, YouTube. https://www.youtube.com/watch?v=ESEFUaEA7kk.

thank the antifascists for forcing the issue and exposing the severity of the antisemitic rot.

Of course one might imagine the editors of the blog in question protesting that they had been merely too lazy to read the submission at all—*fair enough!*—and then directly denounce and repudiate the arguments of the piece. This might repair most of their reputation. But no one disputes that editors of *Social Text* did, in fact, read Sokal's piece—certainly its forthright opening paragraph—and they never directly repudiated its declarations nor admitted that platforming such claims was beyond the pale. Indeed it's quite fashionable to this day for postmodernists to sneer that while some mere *technical details* may be wrong, they actually find nothing major to disagree with in Sokal's piece.

But while his collaborators tried to change the conversation, Aronowitz couldn't bear to give up the fight and, amid trying to cast his team as "*rock and roll*" being attacked by squares, kept returning to direct engagements on antirealism where he would snarl that anything slightly weaker than the most intense strawman of realism was sufficient to derail the realist project entirely.

> "The expectation that physics can generate a theory of the unified field [sic] has definite metatheoretical presuppositions: that the object of inquiry is given and is not itself constructed; that rigorous wrought propositions correspond to a reality entirely independent of the process of inquiry; that prediction is not fetishized intervention; and that method can filter the various influences that motivate and surround scientific inquiry... To say that these views are naive forgets the degree to which Science, despite its secular protestations, has become a religion with many denominations."
>
> Stanley Aronowitz, response to Norman Levitt, *The Cultural Studies Times*, 1997

In return critics of postmodernism, like the ecologist Harold Frumm, explicitly brought up questions of faith healing and the theosophic cult Christian Science, demanding that Aronowitz clarify his description of scientific accounts of reality as merely just "*one story among many stories*." [72]

Aronowitz never responded.

During the height of the Science Wars, between 1994 and 1997, my family had escaped homelessness to an expanse of housing projects in which gangs and violence was commonplace. I was perceived as a starkly feminine boy and this meant regularly being tested in fights. After being stuck with the ambulance bill the first time I ended up passed out in a pool of my own blood, my mother—a fervent believer that physical matter does not exist and *belief*

---

72 Harold Fromm, "My Science Wars," *The Hudson Review*, vol. 49, no. 4 (Winter, 1997): 599-609.

makes "reality"—browbeat the principal into not calling ambulances again but instead personally driving me home whenever I was wounded. This is how I would end up waiting in her car, holding my scalp together while she did chores, before eventually—deeply annoyed that I hadn't simply *believed* my wounds into being closed—driving me to the hospital for the stitches.

In the waiting rooms, popular magazines breathlessly covered the Science Wars.

4

THE ANTIREALIST CONSTELLATION

"Let us guard against saying that there are laws in nature. Knowledge must struggle against a world without order, without connectedness, without form, without beauty, without wisdom, without harmony, and without law. That is the world that knowledge deals with... What assurance is there that knowledge has the ability to truly know the things of the world instead of being indefinite error, illusion, and arbitrariness?"
Michel Foucault (speaking approvingly of Nietzsche)
in *Truth and Juridical Forms*, 1973

"It is the world of words that creates the world of things."
Jacques Lacan, *Écrits*, 1966

"The validity of theoretical propositions in the sciences is in no way affected by factual evidence."
Kenneth Gergen, *Feminist critique of science and the challenge of social epistemology*, 1988

So far we've used terms like "realism," "reality," and "antirealism" rather freely. This has been necessary to set the discursive stage, but the conceptual associations involved are slippery, multifaceted, and often contradictory—which figures like Aronowitz often exploited. It's time to drill into what the fuck any of this means.

Realism, in the context of philosophy, has nothing to do with the common usage of being jaded and pragmatic (i.e. "why can't you radicals just be realistic"), but rather refers to there being a common reality with structures that we can map with greater or lesser accuracy. We might say that "reality" pretty much by definition denotes that which is singular, universal, consistent, coherent, and somewhat knowable. This is a fairly common working assumption of many people in many cultures, and one might go so far as to call it the default framework converged upon by every toddler as the notion of object permanence, but we can bring it into contrast by noting a number of common expressions of antirealism:

1. There is no such thing as a stable or coherent ontological structure called reality.

2. All that exists is "particulars;" perceptions of any broad, common, or universal, structures are fundamentally mistaken.

3. "Reality" has no extraneous solidity but is mentally or socially fungible, like a collective dream determined by social consensus or degree of personal belief.

4. "Reality" has no universality but is patchy, with facts/structures in one region of some kind different from those in another.

5. At most all we have access to is our most raw visceral experience, but cannot begin to extrapolate beyond that to patterns/structures in reality itself.

6. Our capacity to grasp or make claims about the structures of reality is determined by society and is unresolvably incommensurate.

We can also extend to two categories of direct moral obligations to anti-realism:

> 7. It would be better if we *didn't* grasp any regularities to reality.

> 8. Our *language* and social practices shouldn't attempt to refer to reality.

Often a given anti-realist will slide between assertions in different categories, either holding several at once, or moving between them in a loop. In the above Foucault quote where he builds on Nietzsche, there's a shift between first an ontological claim (what there is) to very quickly an epistemic claim (what we can know).

Sometimes the multiplicity of interpretation between poorly worded banalities and provocative woo is deliberate. Take the following quote:

> "I won't deny that there is a law of gravity. I would nevertheless argue that there are no laws in nature, there are only laws in society. Laws are things that men and women make, and that they can change."
> Andrew Ross, lecture at the New York Academy of Sciences, 1996 [73]

The banality is that if you define "law of gravity" as only our current model of gravity and not as referring to the actual physical dynamics captured by that model, then yes, of course, it's possible that our model will change and improve; but this is trivial and does not warrant such a contrived phrasing. Similarly, you can choose to read that passage as saying that there is an object-level regularity described by our account of gravity, but that calling it a "law" is inappropriate. However, the intended reading that those two retreats are shaped around, the deliberate and intentional provocation, is that the regularities pointed to by the inverse square law or the gravitational metric exists only inasmuch as the law against jaywalking can be said to exist.

This kind of slipping back and forth is rampant across thinkers. Nevertheless I think it's edifying to break these categories out. And—just to annoy certain clowns who think philosophy should be more like poetry and declare any use of numbers to be "satanic"—I'm going to refer to these antirealisms by the above numbering.

---

[73] Alan Sokal, "Reply to Fish's NYT Op-Ed," sent to *The New York Times* (unpublished), May 1996, http://jwalsh.net/projects/sokal/articles/skl2fish.html

ANTIREALISM1: THERE ARE NO STRUCTURES TO BE KNOWN

Antirealism1 is a common enough assertion in many subcultures:

> Conceptual structures simply cannot grasp onto anything. Anything you might think you "know" is never even approximately or partially true but completely and absolutely false. Logic is an illusory phantasm, as is existence, persistence, experience....whatever you feel like can be added to this pile. It's not that we merely can't know the world; there's nothing to be known.

This is basically the faith of ontological nihilism. And you might hear it in, say, any given punk house with chaos magick practitioners because it's got some impressively edgy bluster. But it's rarely held consistently, because to do so would terminate all thought and action, since such fundamentally involve persistent structures, even if just for a nanosecond.

The world or whatever cannot be *infinitely* complex in the sense of resisting structure, because if it was there would be no relation, no connection possible between any given points (in any respect). And once you relax from truly infinite difference and allow even the smallest degrees of commonality to tie things together, that's structure.

Instead those who occasionally make such claims in their rhetoric, like Peter Lamborn Wilson, tend to leap back and forth between it and other forms of antirealism, like Antirealism3, where your thoughts can control reality, or to a variant of Antirealism8 in which the point is to undermine the word "real": *Everything is equally "real" including illusions.* It functions as a move in a dance, where less intense antirealism is not sustainable without a retreat to higher skepticism of Antirealism1, but holding "*reality doesn't exist*" is also infeasible and so must be progressively softened after retreating to it. Antirealism1 is thus always just a momentary rhetorical salvo, not an honest ontological commitment.

> "The universe is naturally chaotic. When someone tries to impose order on some small part of it, the order will inevitably come into conflict with the chaotic universe and will start to break down... It is wonderful; it is magickal. It is beyond any definition."
> Wolfi Landstreicher, *Rants, Essays and Polemics of Feral Faun,* 1987

ANTIREALISM2: THERE ARE NO
GENERALS OR UNIVERSALS, JUST PARTICULARS

Antirealism2 is a slightly more relaxed version of the edgy battlecry, a variant more at home in academic polemics than chaos magick zines, but still in practice collapses immediately because, again, *all thought and action requires structure and generalization.*

An assortment of complete particulars could have no relation or commonality between any two of them and thus no capacity for transition between them. Every instant, point of existence, perspective, etc. would be utterly unrelated to every other. And of course, beyond the most bare-bones pragmatic considerations, the assertion that everything is particulars with no universal connections spanning between them, is itself a universal declaration, a form of sweeping structure.

Pretty much every vaguely thoughtful commentator quickly pointed this out to its proponents.

> "Assertion of the pure particularism, independently of any content and of the appeal to a universality transcending it, is a self-defeating enterprise."
> Ernesto Laclau, *Universalism, Particularism and the Question of Identity*, 1992

Both ontological nihilism and absolute particularism describe a frame in which thought is not possible, in much the same way that it might be possible that you are brain whose internal structure (e.g. prior memories) randomly spawned into existence a moment ago and your entire brain will dissolve in another moment, there is nothing that can be thought or done vis-a-vis that hypothesis. No intentionality that we can map onto it or through it. We can't even say that it's "useful" to believe a given thing were such hypotheses the case, because there is either no coherence or no commonality across infinitesimal particulars such that "usefulness" in any sense can reach.

An ostensibly more palatable version of particularism was gestured at by Richard Rorty as "pan-relationism" where there's infinite relations between particulars but each relation is itself utterly particularized in the sense that for some reason there's no generality or structure to the overall network of relations.

Rorty desperately wanted to avoid, in his words "metaphors of depth"—anything that might gesture at root dynamics—in short, he wanted to avoid *radicalism*. What Rorty wanted was a picture of the world that was absolutely "flat," where no particular relation (or relation between relations) could be weighed as more sweepingly important or innately influential than any other. You can bounce around between particularized relations, but there is no "cheating" by finding the meta-relations that compress, span, or connect them.

Beyond the flat absurdity of claiming that a potential poetic relationship of "cookies" to "sunset" is the same importance in the network of all possible relations as the relationship mapped by electromagnetism, the problem is that asserting the absolute flatness of all relations as a truth is itself to posit a truth and a universal meta-relational structure, and an arguably very implausible structure at that. The assertion that there are no root relations is itself a claim of a root relation. This is a trap Rorty was quite aware of:

> "The difficulty faced by a philosopher who, like myself, is sympathetic to this suggestion [e.g., Foucault's]—one who thinks of himself as auxiliary to the poet rather than to the physicist—is to avoid hinting that this suggestion gets something right, that my sort of philosophy corresponds to the way things really are."
> Richard Rorty, *The Contingency of Language*, 1986

Rorty's response was to entirely retreat from the ontological claim of Antirealism2 that there are no universals or generalities, instead merely positing a moral obligation to reject acknowledging roots or common generalities. We'll get to what he thinks the moral stakes are later. For now it's important to recognize that ontological particularism is only ever an argument made in passing, it cannot be an argument dwelled on or asserted universally because the contradictions blow it up.

From the punk houses to the academic seminars, Antirealism2 thus occurs primarily as a salvo of bravado or intended negation of thought. Those who don't retreat to simply moralizing against recognizing any generality, often retreat to the ontological relativism of Antirealism4, wherein enough generality and commonality is admitted to allow for internally-coherent patches of reality.

ANTIREALISM3: REALITY IS MENTALLY FUNGIBLE

Antirealism3 is always a wildly popular claim and builds the base of any populist antirealist movement. Anyone who has had a lucid dream knows the basic idea. It can be summarized as there being nothing beyond credences or motivations. Gravity holds no external rigid hold on you, all that's stopping you from levitating at will is a matter of belief or desire.

> "All those bodies which compose the mighty frame of the world have not any subsistence without a mind."
> George Berkeley, *Principles of Human Knowledge*, 1710

> "There is no universe, it's a human construct... There are no electromagnetic fields or molecules or atoms or waves of particles, these are human constructs. These are modes of knowing and experience in human consciousness."
> Deepak Chopra, lecture at MIT, 2018 [74]

> "The physical world is the dream. When one 'awakens' from the 'dream' of the physical world, one realizes that the dreamer is the cause of the events and the relationships. The out-of-consciousness collective/universal mind is the creator of the world the individual mind experiences. We spend most of our lives in this 'reality dream.'"
> Cynthia Sue Larson, *Reality Shifts*, 1999

Of course, since you demonstrably can't levitate at will—at least when others are around to fact check—explanations have to be given for why. There are basically five common approaches:

a. When you set your credence in something to 1—absolute perfect belief on the scale of 0 to 1—it becomes "real." Thus any impediments to levitating are impediments to setting your credence all the way to 1. You just haven't deluded yourself hard enough. This is the ideological core of most faith healing. Whatever you have *faith* is real then becomes real. Visualize it harder!

b. A slight variation of (a) is not focused on degree of *belief*, but on degree of *desire*. If you want something sufficiently strongly, if you get your ravenous psychic hunger to a certain level, it will become "real." Just look at the shopping catalog more intensely while the prosperity preacher rants and you will have that gaudy diamond ring. Any impediments to levitating are thus the result of your desire to levitate not being

---

[74] "The Nature of Reality – Deepak Chopra at MIT," *The Chopra Well*, 2018, YouTube, 29:45. https://www.youtube.com/watch?v=CHmnPVApfFE.

sufficiently intense.

c. When you de-rigidify your mind, reality in turn becomes de-rigidified. This can be framed as an alternate extreme to (a): instead of setting your credence in levitating to 1, you set your credences on *everything* to 0. Thereupon the supposedly fixed structures of the universe will dissolve before you and anything will be possible.

d. There's a similar alternative of (b) where the very presence of desire is what makes reality get fixed into seemingly solid patterns and structures, so if you sufficiently diminish all your desires, reality will cease to push back against them and become fluid again. So to be able to levitate you must first truly become empty of all desire, including—conveniently—any desire that would lead you to then test whether you can levitate.

e. When there is "local" consensus in something it becomes provisionally "real." Any impediments to you levitating are thus the result of someone else believing (or desiring) the inverse. The only thing that can stop a person with a belief is someone else with an opposing belief. There is never a static fact of the matter beyond the network of minds with beliefs of varying strengths. So if you can't levitate, or you're dying of cancer, it's a person or a group of people that is responsible, because "matter" has no reality of its own.

> "The principle of reality is other people."
> Bruno Latour, *The Pasteurization of France*, 1988

Often there is some kind of combination of these, mixed with some poorly defined account of a particular mechanism. One common variation is to conjoin (b) and (d): physical reality is created by intensity of individual desire for various things as plays out across a society, and the mechanism of this is linguistic: words create the common experience of materiality, and the apparent structures of reality are constructed via symbols imbued with intentionality. Thus if you want to reshape the apparent physical reality, hack the raw wordscape that gave rise to it by turning a few words into a sigil and focus on it while jerking off.

I wish I was joking or that this was confined to a few punks wasting their lives on such chaos magick delusions, but much of the wider United States population has long been enraptured by the notion that reality is fungible. This can look like Elon Musk, who reportedly subscribes to such extreme sophism that he suspects he dreamed his on-again off-again partner Grimes into existence, as she and the rest of us are mere sock puppets of his own mind.[75] But it can also look like lefty yoga spiritualist types furiously haranguing if you make the mistake of talking about rape or racism, because such negative vibes (e.g. belief that such exist in the world) are what's responsible for actualizing such into existence.

In the theosophic cult Christian Science, religious worship is centered around something called "Testimony" in which everyone gathers in a room and one-by-one stands up and talks about examples from the past week in which they allegedly remade reality with their minds. This can look like, "I thought Susan at work hated me, but then I changed my mind and decided that Susan must love me and then later that week she was nicer to me" where the mechanism (being nicer to someone as a result of your newly adopted attitude) bootstraps a change in a social relationship. But it often looks like "I was in an accident and thought my legs were shattered, but then I believed that it's impossible for anyone to be injured and then the mess of flesh and bone knitted itself back together!" If everyone else has these experiences through their goodness, you are pressured into likewise recounting having such superpowers. Whenever you have a momentary paranoia or delusion—that then evaporates—you are rewarded, until this becomes a habit of getting "mortally wounded" and then "healed by prayer." The approval of your peers habituates such flights from reality, and, when you actually get cancer or break your leg in the real physical world, shame keeps you from acknowledging it. As with Prosperity Gospel, misfortune is simply the result of insufficient belief. Thus when the children of Christian Scientists get sick or even seriously injured there are strong pressures to take them to a registered faith healer (a "practitioner" in the lingo) rather than a hospital.

While "Christian Science" is more derived from theosophy than christianity, the influential christian theologian George Berkeley also infamously rejected the existence of matter, arguing that all we have true access to is our perceptions, so we should reject any attempt to think of or search for "abstract" underlying roots beneath them like, you know, persistent material objects. In Berkeley's scheme—shared by Deepak Chopra[76]—the persistence of material structures against our wishes or beliefs lies in the fact that a singular unseen individual of immense power—a "god" in the parlance of christians—is

---

75 Jade Biggs, "Grimes says Elon Musk thought she was a "simulation" that he'd created," *Cosmopolitan*, October 17, 2022, https://www.cosmopolitan.com/uk/entertainment/a41637053/elon-musk-grimes-simulation/.

76 Deepak Chopra et al, "Why a Mental Universe Is the "Real" Reality," *Chopra Foundation*, https://choprafoundation.org/consciousness/why-a-mental-universe-is-the-real-reality/.

dreaming them into being. The posited infinite strength of this invisible individual in contrast to all other minds thus explains why faith healing often fails.

> "When in broad daylight I open my eyes, it is not in my power to choose whether I shall see or no, or to determine what particular objects shall present themselves to my view; and so likewise as to the hearing and other senses; the ideas imprinted on them are not creatures of my will. There is therefore some other Will or Spirit that produces them."
> George Berkeley, *Principles of Human Knowledge,* 1710

Broadly speaking such belief that the physical universe is merely the product of mind is not exclusively a "western" phenomenon, you find this kind of antirealism (to varying degrees) in many other cultures, including some animist traditions and large swathes of hinduism where faith healing remains a prominent practice. And it's hardly irrelevant to postmodernism. Lamborn Wilson, for example, implied that the collective dreamtime of Australian aborigines (or rather his muddled colonial impression of certain accounts) is prior and more primordial than physical reality.

It's impossible to overstate how wildly popular this was in the Left. To take just one example from a prominent feminist theorist and activist who had spoken before the 1980 Democratic National Convention,

> "In the last decade or so, people in the West have been waking again to the role of the imagination in creating reality. The knowledge of this power, as old as human consciousness, had all but disappeared in those parts of the world where analytical, 'rational' modes of coping gained ascendancy several hundred years ago… But in recent years… the idea that the image precedes the actuality, and in fact gives birth to it, has seeped back into awareness… We had to dream the dream so powerfully and with such faith."
> Sonia Johnson, *Going Out Of Our Minds,* 1987

The facile-level justification is that self-delusion imbues confidence which can sometimes shift probabilities—whatever wild risks and unethical abdication of diligence that might entail—but magick practitioners rarely content themselves with such a limited and pedestrian claim. If magick were about nothing more than unwarranted confidence you could purchase it far more conveniently at the liquor store. The substantive claim is always that through the equivalence, permeability, or non-dichotomy of thought (in representation, signifier, etc.) and material reality, one can simply twist one's thoughts and remake reality.

Johnson herself quite explicitly meant dreaming patriarchy out of existence, and this was inclusive of "hoping" to such an extent as to literally remake history and erase the invention of nuclear bombs. This followed, she argued because "*There are no natural laws, just habits,*" and "*There is no objective reality. Reality is what we accept is real.*" Since reality was "constructed" as a shared dream and "*all systems are internal systems, patriarchy does not have a separate existence outside us,*" the physical subjugation of women was therefore a product of women being tricked into believing it into reality. Amid encouraging feminists to create "*bubbles of psychic power,*" Johnson appealed heavily to the New Age "morphogenic field," a concept that Sokal would later make part of his hoax.

In the '80s and '90s, Starhawk was a very popular writer, one of the few "anarchists" friendly with postmodernists, and her accounts of "manifesting" are broadly indicative of the genre:

> "When energy is directed into the images we visualize, it gradually manifests physical form and takes shape in the material world."
>
> Starhawk, *The Spiral Dance*, 1979

Of course one must immediately caveat that this process is not simple:

> "Casting a spell is like sailing a boat. We must take into account the currents—which are our own unconscious motivations, our desires and emotions, our patterns of actions, and the cumulative results of all our past actions. The currents are also the broader social, economic, and political forces that surround us. The winds that fill our sails are the forces of time and climate and season; the tides of the planets, the moon, and the sun."
>
> Starhawk, *The Spiral Dance*, 1979

Putting aside the necessary avoidance of falsifiable claims, it's important to note that this staunch refusal of conceptual radicalism (that is to say a reduction to clarity, concision, and seeking to "carve reality at its joints") means an embrace of open-ended games of association. Holism thus provides one with poetic license. One that superficially feels freeing, albeit in the same way that daydreaming feels to a prisoner. Nothing changes for the better as a result of this self-delusion, but the framework provides the *sensation* of an increase in freedom.

In anti-reductionist magical thinking the signifier (a word) is—in some deep sense—the signified (the physical object it refers to), so a scrawled symbol becomes what it represents and operations on the symbol become operations on the signified.

> "What we name must answer to us; we can shape it"
>
> Starhawk, *Dreaming the Dark*, 1982

Premodern magical systems often thought that physical reality in one sense or another was language, and thus could be hacked or remade by language. When some structuralists and poststructuralists argued—from the other direction—that everything was language, it should be no surprise that the audience most enthusiastically receptive of the emphasis on metaphor and symbolic conflation were those trying to resuscitate magical thinking.

But of course, not everyone attracted to postmodernism were magical kooks. Some were more conventional reactionaries.

ANTIREALISM4: REALITY IS PATCHY

Antirealism4 might be called "ontological relativism." Most people embrace it in hopes of bolstering a commitment to moral relativism. In this sort of approach the very ontology of reality is highly relative between what might be termed "patches." These can be spatio-temporal patches, but they can also be patches within a social landscape or patches of different levels of scale.

This sort of thing is most typically appealed to by the currents of fascists and nationalists who claim the banner of postmodernism and leverage its arguments primarily against universalism. Much of Nanda's career has been focused on exposing this dynamic in the case of hindufascists screaming that Indian science is ontologically distinct.

> "The Indic tradition largely accepts Western science as valid within its own sphere, but regards that sphere as limited."
> David Frawley (Vamadeva Shastri), *Awaken Bharat*, 1998

The founder of Christian Science, Mary Baker Eddy, cited multiple passages from hindu scripture in the 16th through 49th editions of the faith's central text. These were later removed as Eddy tried to burnish the cult's aesthetic presentation as christian and in an attempt at distinguishing the cult from hinduism she emphasized the centralization of "god" in the construction of deep reality more in line with Berkeleyian idealism, rather than decentralized creation centered around individuals.[77] Nevertheless, Christian Science has always blamed the stubborn persistence of physical reality and misfortunes on other individuals having conflicting beliefs in a way that very rapidly ends up reproducing a patchy notion of reality.

When a child gets severely injured, it's common for a Christian Scientist to tell them to *commit* to one belief system or another, disbelief in matter or belief in it, but not to mix them, because the beliefs will conflict in creating one's local reality. The doctor or physicist is thus perceived as a kind of magician, in league with their collaborators, whose collective strength of belief in matter, creates and conditions the world. Those who integrate too closely to the materialist patch become slaves to the rules they have believed into existence. It should not be surprising then that functional belief in and fear of "witchcraft" is deeply embedded in Christian Science, since there are loci of belief-power in the world beyond the materialism of scientists, indeed ultimately they believe

---

[77] "What did Eddy say about Eastern thought systems?," *the Mary Baker Eddy Library*, September 18, 2023, https://www.marybakereddylibrary.org/research/what-did-eddy-say-about-eastern-thought-systems/.

each individual generates a reality patch to some degree. I once brought an occult goth I was dating home to dinner with my mother for shits and giggles and the two of them stared wildly at each other the whole time, each explicitly trying to will the triumph of their local reality patch against the other's.

But another example of "patchy reality" that has grown very prominent is the fascist relativism of Alexander Dugin (who merely repackages common currents from Martin Heidegger to Ivan Ilyin). In Dugin's far more collectivist and nationalist approach, insofar as one is within the geist of "Eurasia," a fact like, say, the speed of light, could be everywhere in the universe 100 million m/s, whereas insofar as one is within the geist of the "atlanticism" it would also ontologically be the case that the speed of light just is 299 million m/s everywhere within the universe. Two different ontologies and both are simultaneously correct to infinite spatial extent. Of course these patches, or alternative universes, interface somehow, as we clearly can talk to, shoot bullets at or even convert between geist patches but don't you worry about how commutativity functions in moving between these patches since logic itself follows this patchiness.

It's important to understand that this is *alethic* relativism, not merely epistemic relativism: what *is* the case varies, not merely what is *perceived* as the case. The law of the excluded middle happens in one patch and not in another patch, and this doesn't cause problems somehow because the patches just vibe together. You can't apply a meta logic system across all the combined patches as this would undermine the entire idea. Thus there is no singular unified reality. Latour, for example, smirks that what universal law of physics applies in Paris should not be assumed to apply in Timbuktu.[78] When the patches follow the contours of nations you get "Indian physics" or "Aryan physics" but when the patches are individual level something like "Dave pulled his finger on the trigger of the gun that shot Emma" may be personally true for one person but false for another person, and each "reality" is just as real or factual. If you've been a part of any radical subculture and haven't encountered numerous people smugly appealing to such alethic relativism in instances of rape or violent abuse, count yourself lucky.

For a more academically prestigious and less ostensibly political example of Antirealism4, consider certain rejections of scientific unification (e.g. the entire reductionist or radical project of science), wherein it is asserted that the domain of structures or regularities in nature studied by biology do not reduce to that of chemistry and then to that of physics but rather somehow constitute independent ontological patches. These patches might be able to

---

78 Bruno Latour, *The Pasteurization of France*, trans. Alan Sheridan and John Law (1998), 226

communicate or interface at their mutual borders, but there is some kind of magical force at the ontological level that prevents them from simply being different descriptive levels of the same thing. These patches supposedly don't reflect differences in description of a single reality, nor differences in precision or capacity for fidelity, but differences in reality itself.

Biology, in such an approach, isn't a system of rough generalities and rules of thumb describing imperfect patterns abstracting over particle or molecular dynamics—a usefully abstract coarse-graining that can't violate the predictions of the more fundamental physics account—but somehow an entirely different thing, constituting different root causes. This a distinction between saying that our concepts of "hand" or "lion" are pragmatically useful practical abstractions with some inescapable fuzziness and saying that some "hand-ness" or "lion-ness" beyond the causation covered by the laws of physics imposes itself down on the constituent atoms. Thus even if physics, for example, describes absolute limits to how energy flows, "biology" can—even if this has never been demonstrated—somehow override or contradict such.

This dream of antireductionism and "top-down causation" (as opposed to the normal "bottom-up" kind) has a long history with regard to biology specifically. The French philosopher Henri Bergson, who infamously embarrassed himself debating Einstein, was a huge proponent of "vitalism"—the notion that "living beings" are powered by some kind of mystical non-physical force—and Bergson was still quite influential in the Parisian scene of the '60s and '70s. In leftist circles there—it's hard to call them "radical" since the rejection of bottom-up accounts of reality is by definition a rejection of "radicalism"—currents like the journal *Tiqqun* continued to militantly stump for "vitalism" even into the '00s, with *Tiqqun* founder Julien Coupat even extending into COVID conspiracism.[79]

Similar antireductionism found popular root in the American mystical left, where the claim of causal patchiness to reality to stop physics reduction was seen as both morally necessary and a pathway up to justifying the magical dynamics of Antirealism3.

> "Vitalism in its monistic form was inherently anti-exploitative. Its emphasis on the life of all things as gradations of soul, its lack of a separate distinction between matter and spirit, its principle of an immanent activity permeating nature, and its reverence for the nurturing power of the earth endowed it with an ethic of the inherent worth of everything alive. Contained within the conceptual structure of vitalism was a normative constraint."
>
> Carolyn Merchant, *The Death of Nature,* 1980

[79] Douglas Morrey, "Manifeste conspirationniste, Parti imaginaire, Comité invisible: a genealogy of radical critique in twenty-first-century France," *Modern & Contemporary France,* vol. 33, no. 2 (2025): 209-224.

* * *

We might call all the antirealisms discussed so far ontological antirealisms. Antirealism1 claims that there is no reality whatsoever in any sense. Antirealism2 claims there is no commonality. Antirealism3 says there is no fixedness or consistency independent of us. Antirealism4 says there is no universality.

In contrast the second batch of antirealisms are epistemic and linguistic—they concern themselves less with claims about what is than with claims about what we can know and what we can say—although they often fall back on, implicitly require, provide cover for, or rapidly collapse into the previous antirealisms.

ANTIREALISM5: KNOWLEDGE OF REALITY IS ENTIRELY IMPOSSIBLE

> "Absolutely nothing can be predicated with any real certainty as to the 'true nature of things'"
> Peter Lamborn Wilson (as Hakim Bey), *Immediatism*, 1992

Antirealism5 emphasizes certain impediments to knowledge of the structures of reality, inflating small frictions to total barriers. This sort of antirealism takes trivial dynamics—like that our initial perspective can have an influence on the patterns we pick out, or that we have limited intellectual resources, or that different models can describe the same evidence—and inflates them into absolute barriers to all knowledge of any degree.

Four common lines of attack here are:

a. Emphasizing the network nature of language until it may seem reasonable to imply that our concepts can't refer to *anything* outside human discourse.

b. Saying that because observers are not entirely passive receptacles of structure we completely build our models of the physical universe without any substantive regard for it.

c. Arguing that because there is always an infinite number of models that can reproduce any finite set of data, we have no way of deciding between theories.

d. Just outright imperiously declaring that there is no universalism or unity to our models, no single elephant to be found underlying different investigative projects.

We've covered how the French structuralists inherited an anti-reductionist perspective that treated everything as merely one part of a whole. In the structuralist paradigm the signifier "red" is not defined by what is signified—some relation to the physical world like a wavelength range—but instead by not being other potential and actual signifiers like "dddddd," "triceratops" and "blue." When you add in the bald assertion that our concepts are shaped entirely by that language network, all notion of a referent is sliced away. Everything becomes purely a matter of language and there's no place left for causal lines outside the social world, like individual experience or sensory engagement with matter. This is related to what Quentin Meillassoux calls the Correlational Circle, a now common maneuver that Berkeley popularized:

the argument that we can't think about a thing "outside of thought" because to think about it is to still be only thinking about it. We'll cover the impulse underneath this more in Antirealism8.

But the notion that we can't in any way access the structures of reality, takes a variety of other forms.

Of greatest influence in the western tradition of philosophy was Immanual Kant's infamous anti-copernican revolution, a desperate response to the humbling discoveries of science that left humans as incidental rather than central to reality. In Kant's approach, humanity was re-throned as the center of the epistemic universe:

> "Hitherto it has been assumed that all our knowledge must conform to objects. But all attempts to extend our knowledge of objects by establishing something in regard to them a priori, by means of concepts, have, on this assumption, ended in failure. We must therefore make trial whether we may not have more success in the tasks of metaphysics if we suppose that objects must conform to our knowledge."
>
> Immanuel Kant, *The Critique of Pure Reason*, 1787

The core argument is that we supposedly can't know something without having some pre-existing foundational conceptual framework to form further concepts within and who can say how much that reflects any actual reality. In an underhanded linguistic flip, the postmodernists would make all the rage and that we will cover when discussing Antirealism8, Kant also implicitly shifts the definition being used of "object" to merely our concept. Infamously Kant applied all this to Euclidean space, considering it as an exemplar of a fixed preconception all minds magically have *before any experience*:

> "Space is not something objective and real... it is subjective and ideal, and originates from the mind's nature in accord with a stable law as a scheme, as it were, for coordinating everything sensed externally."
>
> Immanuel Kant, *Inaugural Dissertation*, 1770

That such a howler was promptly disproven by scientific discoveries that radically revised our notions of spacetime has had little impact upon kantians and the philosophical currents that respected him. Despite his intention to push back against the more intense idealism of folks like Berkeley, at the end of the day what Kant provided to many was reactionary catharsis to the troubling conceptual revolutions wrought by physicalism. Oh sure, do your science or whatever, but know that models don't discover or grasp the world as it is, they all merely derive from our starting perspective.

Of course this move to emphasize subjective origins was surely stale when

Protagoras made it 2500 years ago, but in Kant's historical moment it arrived to significant acclaim. And it should perhaps be unsurprising that a certain type has consistently fallen back on this perspective for the comforting reassurance that there is no cold hard impersonal reality whose structures and regularities press incessantly on us, shaping, containing, and determining our conclusions. Social realities can be negotiated, persuaded, beguiled, and overthrown, and many who operate primarily in such a realm instinctively want to declare it unbounded, even by mathematics and logic.

> "We think the world is logical because we have made it logical."
> James Leigh, translating/endorsing Nietzsche, *Semiotext(e)*, 1977 [80]

What you will immediately note about Kant's notorious argument is that it assumes a stark remove into existence. Why would it necessarily matter what initial conceptual framework we start with? Even if Kant had been right that all minds necessarily start with an implicit conceptual frame of euclidean space, we demonstrably can revise such starting frameworks. But Kant inherited a cleaving of the world into that which can be known as disembodied minds prior to engagement with matter, the *a priori*, and that which is known as a consequence of engagement, the *a posteriori*.

In contrast, if we are already inherently from the start embedded in and a part of the world, then there can be no firmly fixed lens through which we view everything. Every possible lens is subject to revision from the pressure of the object upon the subjective. Thus we think the world is logical, not because we invented logic truly from nothing, but because the world has beaten logic into us, forced us to change ourselves until we had fashioned our brains into what could clasp onto the fixtures of the world. Starting points don't matter, knowledge arises from engagement. We start off thinking the sun revolves around us and spacetime is flat but then update our account. At least we do if we're not Kant.

It's important to note that Kant saw himself as *resisting* the antirealism of Berkeley. Kant emphatically endorsed the notion that there was an exterior physical universe, something firm and not dreamed into being and constantly shaped by someone's thoughts.

Yet his emphasis on the supposedly totalizing grip of our own subjectivity meant that this *reality-in-itself* was an utterly unknowable thing. It *somehow* influences our perceptions and concepts but never in any way that could allow for our ability to reference *reality-in-itself*.

In this view, our growing appreciation for the curvature of spacetime does

---

[80] James Leigh, "Free Nietzsche," *Semiotext(e)* III no. 1 (1977): 4-6.

not reflect an actual reality outside of us that our thoughts are latching onto. Our models cannot press ever more closely against the skin of reality until any gap remaining is so miniscule as to be only a technicality, or even to the point where we can say that an impression of ours merely reflects a structure that exists. We are thus stuck forever referencing exclusively shadow puppets cast by our own mind.

> "In a certain sense all perception is projection."
> Theodor Adorno and Max Horkheimer,
> *The Dialectic Of Enlightenment*, 1944

> "To the extent that the mind furnishes the categories of understanding, there are no real world objects of study"
> Kenneth Gergen, *Social Psychology as History*, 1973

As with many academic subjects where there's entire industries of exegesis, it's unclear precisely where Kant's personal emphases would shake out or how he would shift if he were whisked through time and introduced to things like general relativity or neuroscience, but many currents have certainly taken from him the impression that it was authoritatively demonstrated that we simply cannot know anything *at all* about the external physical world.

From the beginning, philosophers have made the obvious snark that this sort of argument is just antirealism plus a free-floating assertion of somehow being a realist nonetheless. *"Fig-leaf"* realism. You get to defensively pronounce that you're a realist, but to absolutely zero impact. Very much akin to how liberals will declare that they're *opposed* to fascism, but steadfastly reject fighting fascists or taking literally any action whatsoever. They'll turn you over to the cops, be compliant to the fascist regime, denounce antifascists, endorse fascist arguments, etc.. In short, all their opposition to fascism amounts to is additionally reserving the right to, under specific circumstances—like a dinner party—utter the phrase "I'm opposed to fascism."

Similarly, in the Science Wars, there was a faction that emphatically asserted that they were realists, but only in a sense of the term where "reality" didn't have any constraining effect on our models of it. This faction of sociologists would declare themselves "naturalists" and claimed to believe in a physical material universe, but would fight to the death against any implication whatsoever that the structures of this reality could play a role in determining our models of them. As these sociologists developed their positions under continuous conflict with philosophers and scientists, they pulled primarily from formal analytic philosophy rather than postmodernism.

Figures like Barry Barnes and David Bloor, embraced the term "relativist" while explicitly appealing to Kant's conception in their self-description:

> "Relativism with a realist flavour or realism in its most attenuated form, sometimes described as 'thing-in-itself' realism."
>
> Barry Barnes, *Realism, Relativism, Finitism*, 1992

But whereas Kant had emphasized our subjective *starting point*, this clique of sociologists leveraged Kuhn and Feyerabend to emphasize the subjective lens involved *every step of the way*, as our existing theories inform how we search for and parse new data.

Every time we evaluate data we do so from the perspective of existing theories or conceptual schema. For example, we might recognize some data as valid disproof of a theory, but be so locked into a perspective that we don't even see other conflicting data, dismissing it as background noise or experimental error.

This dynamic of the *theory-ladenness* of our perceptions and descriptions has long been a known friction in the improvement of scientific models, but the relativist sociologists emphasized this dynamic not as a matter of degree, but as a binary, a poison whose tiniest presence catastrophically and absolutely invalidates every model we might build of the world.

To do this, they had to reject that reality in-itself could have any impact on which models congealed and which disintegrated. There could be no feedback loops of engagement by which theory-ladenness became suppressed as a problem, where lurking faulty priors could be identified and corrected.

Functionally they had to side with those who declare all pattern extrapolation whatsoever to be bunk.

This is usually talked about in terms of "induction" and is almost always presented *via* some toy cases that leverage your intuitions. That all swans you've encountered have been white doesn't mean the next one can't be black. Just because rocks have fallen to the ground before, doesn't absolutely prove they'll fall to the ground tomorrow. Just because the Sun has risen all your life doesn't mean it will tomorrow, it might go off and do a jig in another galaxy instead.

There's a sense in which this is fair. When you encounter 1, 2, 3, 4, 5, 6 at the beginning of a mathematical sequence this might appear to be a pattern of the integers, generated by a simple "+1" operation where the next two entries are 7 and 8, but it's certainly possible for the underlying formula generating the sequence to be instead arbitrarily discontinuous, weirdly gerrymandered to abruptly transition to something very different past a certain number of terms. And one doesn't always need crude discontinuity, complexity above the simplest extrapolation can be introduced in other ways: the Doudna sequence[81] goes 1, 2, 3, 4, 5, 6 before suddenly spitting out 9 and then 8. Similarly, if the underlying root mechanism generating the sequence of numbers is "*k such that (8*10 ^{k} + 49)/3 is prime,*" well that sequence goes 1, 2, 3, 4, 5, 6 before spitting out 10 and then 24.

It is important to point out that the standard ways of presenting *"the problem of induction"* is in terms of linear sequences, but reality is not linear, it is intensely multidimensional, with relations and structures massively interrelating with one another.

Our intuitions often shift when, instead of linear math sequences, we are presented with a photo or puzzle with a few missing pieces, but it's the exact same issue of induction or pattern extrapolation.[82] What's going on is that an increase in dimensionality enables an increase in lines of confirmation. And this points to why we feel greater firmness in predicting the Sun will rise than all swans will be white. The earth's rotation in relation to the Sun is confirmed to us *via* far more avenues of inductive pattern, building on one another in a web far more complicated than "all the dots in a sequence have been white so far so they'll be white in the future."

Often in science (and in life) the question is not one of *prediction* but one of *compression*, because you more or less have all the data in a domain you'll ever get, further experimentation is too expensive, the distant horizons unreachable, but the search for knowledge of course continues. You have to search for and evaluate between different possible underlying patterns, generative rules, etc, that can explain or compress the data that you have.

But of course all data will be finite, and there are, technically, infinite possible underlying structures that can give rise to any finite expanse of data—*as long as you're okay with the complexity of these postulated structures going to infinity very fast.*

Put in other terms, yes, you *can* find underlying mechanisms other than "+1" that generate the sequence 1, 2, 3, 4, 5, 6 but there's very few that aren't severely messy and convoluted.

If theory construction is seen as mere random grasping in a vast void of infinite potential underlying structures, an infinite number with infinitely arbitrary complexity, there's nothing stopping there from being far better theories lurking, like a lovecraftian god, far beyond our capacity to grasp. In

---

81 Currently A005940 on the On-Line Encyclopedia of Integer Sequences

82 From this point in the text I'm going to continue using "induction" as a placeholder for a broad category of inference, inclusive of abduction, etc, because 1) many of these antirealists hounding on "the problem of induction" do so, and 2) I think distinctions between rule and cause are unsustainable in much the same way as between relation and relata.

this sense, skeptics remind us that our best theories may simply be the *"best of a bad lot."*

Johannes Kepler is famous for eventually grasping that the planets travel in ellipses, but his first theory was that they traveled along the surface of six giant spheres separated by the five platonic solids. By sheer coincidence this got the ratio of planetary orbits surprisingly somewhat right. Kepler eventually realized he'd gotten it wrong on his own, and the discovery of Uranus would definitely have derailed it, yet he got *close* by happenstance.

The relativists claimed that even if our models did perfectly predict the future, that's only to be expected from a "mere" whittling down of our theories. We may simply have latched onto a model that correctly predicts everything but yet somehow has no relation whatsoever to the actual structures of reality.

> "There is no valid deductive pathway from 'theory T works' to 'theory T is true.' The reason is simple: false theories can make true predictions."
>
> Barry Barnes, David Bloor, John Henry,
> *Scientific Knowledge*, 1996

Of course what is avoided in such arguments is *probability* and *structural content*. How likely are we to find false theories that somehow perfectly match every testable prediction the true theory would give? And does such commonality force the false theory into sharing significant structural content with the true theory?

Kepler's simple platonic solids theory stumbled into figures within spitting distance, but this stops seeming so strange when you consider he was only trying to predict *five* values in a linear sequence, and "spitting distance" meant deviations of almost 9%. Modern science is so many orders of magnitude richer and vaster, and this dramatically changes the likelihood and character of alternative theories.

Is it *just as likely* that the underlying structures of the world are infinitely complex as simple?

It is at this juncture that antirealists often reach for an escape hatch: they try to leverage induction against itself.

Since we have found patterns in the past that eventually failed to hold up, all patterns we perceive now must also eventually fail. Since some scientific theories have catastrophically failed in the past, *all* scientific theories in the present must encapsulate absolutely zero degree of reference to the structures of reality.

> "Since all theories and other leading ideas of scientific history have, so far, been shown to be false and unacceptable, so surely will any theories that we

expound today."
>> Yvonna Lincoln and Egon Guba, *Naturalistic Inquiry*, 1985

To this day, I very sharply remember a friend in high school complaining, as we worked together in chemistry, that there was no point in learning the material because past theories had been disproven so surely everything we were learning would be disproven too. The structure of compound molecules? Surely just confused nonsense. The existence of particles like hydrogen? Certain to be laughed at in a couple hundred years.

And it's only a short hop from there to think that not just the *entire* content of every current scientific theory will be utterly invalidated, but that all approaches and *strategies* converged on by scientists will as well.

> "What science claims to know is based on a vast body of unexamined assumptions about the nature of language and the language of nature. Even on its own terms it's worth asking, if all scientific paradigms tend to wear out with time, why shouldn't science as a whole do the same."
>> David Watson, *Against The Megamachine*, 1998

It's worth dwelling on how *weird* this attack on our pattern-finding is.

Objects fall to the ground, and an infant develops the scientific theory that this will continue tomorrow. Has *that* scientific theory really ever been disproven? Our theoretical account of falling has deepened and widened, to be sure, but apples, when broken from branches, still fall to the ground. We have made small missteps and corrections in deepening our account of falling objects, but certainly nothing *100%* catastrophic.

Much of the same initial scientific pattern we extrapolate as children persists into Newtonian mechanics, just as much of the structural content of Newtonian mechanics persists into general relativity. We take a step or two backward when we correct and replace a theory, but there are usually more steps forward than back.

Who can remember all the false theories the infant might've constructed first? Why should their initial pratfalls matter to the point where everything after is invalidated? Why should initial stumbles forever overrule the massive interlocking mesh of confirmation that continues forever from that point? And why must there be a temporal asymmetry and privileging of the past? Once upon a time the resources devoted to science were very small, today we devote a much larger global network of diverse attention and much greater resources. If the entire content of scientific models of reality were structurally unsound, if there was *zero* truth to even any chunk of it, we would expect to see regular massive collapses and wild revisions that leave nothing behind, as more minds and experiments were turned on it. Yet we see the opposite.

Surely a theory's stability through the 20th century should count as a much *longer* period of confirmation than a prior theory's stability through the 18th because more iterations of scientific investigation took place in a century with many orders of magnitude more scientists, and vastly expanded access to the universe.[83]

And why is the induction used for science guaranteed to always be proven faulty, whereas the induction used to assert the inevitability of said faultiness is beyond reproach?

Yes, scientists regularly engage with limited data, competing theories, and the potential of unknown unknowns. But they also deal with many subjects in which the evidence is overwhelming and comes from every possible angle, fencing in a question until there is only one possible conclusion. This is why scientific realists often draw on the analogy of a crossword puzzle, emphasizing examples where what we construct is forced by the combination of boundary conditions and internal consistency towards a fixed conclusion.

> "Geologists, chemists, biologists and many physicists tend to be impatient when they hear about the problem of the underdetermination of theory by evidence. A common response is to declare that this is simply a philosopher's problem (in the pejorative sense), a conundrum that people with a certain quirky intelligence might play with, but something of no relevance to the sciences."
> Philip Kircher, *Science, Truth, Democracy*, 2001

What all scientists—and almost all humans—reject out of hand from the get-go is the notion of Antirealism1 and Antirealism2, that the world is just an arbitrary mess, without any underlying structure at all. But excluding these sorts of possibilities necessarily puts *bounds* on the potential complexity of any underlying structures, implying in at least some ways that less compressive models of the world are less likely to be true.[84]

In contrast, the relativist sociologists never recognized any objective epistemic pressure against arbitrarily complex justifications. They repeatedly emphasized that arbitrary discontinuous tweaks and patches could be applied to any existing model to match any physical evidence, thus physical evidence was surely of zero importance in influencing theory. In this sense—because you can add apologetic epicycles, however arbitrarily contrived and awkward, to any wrong theory to make it fit data—all our theories are thus *"underdetermined"* by data.

---

83 Sherri Roush, "Optimism about the Pessimistic Induction" in *New Waves in Philosophy of Science*, ed. PD Magnus and Jacob Busch (2010)

84 Modulo some nuances about the breadth of one's model and assuming non contrived (themselves arbitrarily complex) probability distributions over complexity.

> "Even if empirical input could provide reason for change it could never be considered to provide sufficient reason—unless, that is, the Duhem-Quine hypothesis were to be cast aside... Existing knowledge can always be sustained against new experience without doing violence to reason."
>
> Barry Barnes, *Realism, Relativism, Finitism*, 1992

> "Formally speaking, anything goes. Far from telling us anything in particular, current experience can be described as consistent with any extant body of ancestral knowledge, even offering inductive confirmation of it. Equally, it can be made out as refuting any specific body of knowledge or item thereof."
>
> Barry Barnes, David Bloor, John Henry, *Scientific Knowledge, A Sociological Analysis*, 1996

> "If we take the thesis of underdetermination of theories seriously, relativism is a consequence that is inescapable in some form."
>
> Mary Hesse, *Revolutions and Reconstructions in the Philosophy of Science*, 1980

It's important to note that the relativism being embraced here is not a *cultural relativism* between known human societies, but a relativism between pretty much all existing cognition (which depends upon induction/compression and the assumption that Antirealism1 and Antirealism2 are false) and some hypothesized perspective of a fictional "mind" in which the comparative accuracy and simplicity of theories would count for nothing.

In this perspective, compressive simplicity is a pragmatic virtue *only*. And it's certainly true that we humans have limited minds and practical aims, thus we desire limited accounts of the world in order to operate in it. We could not swing a hammer if we had to mentally track every particle in it and interacting with it. Instead, we collapse the vast complexity of the metal, air, and wood into a crude picture of "hammer" and "not hammer," yet this conceptual slicing up of nature is obviously not in any way objective or obligatory, reality itself has no notion of a "hammer."

It's reasonable that sociologists think this way, after all there's no such deeply objective thing as "capitalism" or "Disney"—an alien mind of cosmically different scale visiting the Earth might not recognize humanity as a distinct "species" or even a distinct phenomenon from the rest of the biosphere, much less make the same distinctions or clusterings of phenomenon within our society. Even when our embodied constraints are the same, different cultures will, for example, slice up the color spectrum in different ways and generate different color theories, tailored to the different working mediums their artists use.

There is no notion of anything like an *objective* algorithmic complexity in everyday human life. In our normal experience, complexity is always profoundly relative to something else. What is simple for one person towards a given end may be complex for a different person towards a different end.

This is especially the case in sociology.

Rarely is sociological data so overwhelming as to confine or studiously box in one's theory to only one conclusion (or really any truly limited subset). In a lot of the social fields, there's always an out, there's always a lateral move one can make in theory to escape some constraint, so folks assume the same principle must apply when inert matter is involved.

Instead of the crossword example, folks operating in this domain would perhaps be more inclined to see *every* sort of speculation—from sociology to fundamental physics—as a photograph in which almost all of the center is missing. If you can see only the outer borders, you could try to extrapolate inward, but there's a lot you could get wrong. You might continue a tree line from one side to another and miss that there was supposed to be a mountain in between. The boundaries of the photo are not tightly coupled with the content and there is no real pressure to not stipulate arbitrary content. There might be a person in the center, multiple people, an animal, a house, a falling leaf, a UFO, etc. So many options as to make it hard to map even the extent of possible theories. The boundary conditions would so wildly underdetermine the content that speculation might seem absurdly unlikely to have any traction on reality. Such speculation might be fine enough exercise for self-amusement, or even serve some darker social ends (who might be chosen to replace the comrade erased by Stalin?), but it would at most reveal the biases and baggage of those speculating.

Aronowitz, for example, badly riffs on Quine in discussing fundamental physics,

> "'Underdetermination' by boundary conditions [eg material experience] points to the proposition that correspondence, even of the 'corporate body' of scientific statements with a physical world that is known by means of experience and which is capable of refutation, is improbable. ... Even if experience is a real boundary limiting the arbitrariness of reference, the idea of underdetermination indicates the function of 'experience' is to force reevaluation but cannot determine the nature of the adjustments. ... Hence, for the most part, science is self-referential inquiry. Either it refers to its own logical presuppositions, themselves man-made, or to extralogical influences that may be designated as 'cultural,' 'political/ideological,' or whatever."
>
> Stanley Aronowitz, *Science As Power*, 1988

This was put in similarly extreme terms by Kuhn in one of his more anti-realist moments:

"Is it not possible, or perhaps even likely, that contemporary scientists know less of what there is to know about their world than the scientists of the eighteenth century knew of theirs? Scientific theories, it must be remembered, attach to nature only here and there. Are the interstices between those points of attachment now larger and more numerous than ever before?"
Thomas Kuhn, *Logic Of Discovery Or Psychology Of Research?*, 1970

Putting aside the vast wealth of experimental "points of attachment" with nature in our modern era of particle accelerators, it's certainly true that theorizing *can* introduce error, but it does not in any way follow that more extensive theorizing beyond base experimentation or phenomenology is innately more suspect! Rigorous exploration and checking can wean down the space of plausible theories. Further, as theoretical depth is developed it can discover vast and concisely interlocking networks of confirmation between distant experimental or phenomenological "points" that would be totally inaccessible without being willing to rigorously explore implications and extensions through theory space.

Because there *are* cases more like a crossword puzzle than a photograph with a hole where boundaries entirely (or almost entirely) determine the content.

In 2011, an experiment between a physics lab in Switzerland and one in Italy appeared to detect neutrinos traveling slightly faster than the speed of light. It's impossible to describe in less than the volume of a book the vast array of things in modern physics confirming or implying the impossibility of this, the sheer *ad hoc* complexity of the fixes to understanding of the world at every level that would be necessary. So the experimental apparatuses were checked, and then checked again. No mistake or error could be found. Theorists took the snowday-like excuse to have some fun writing hundreds of papers on a lark, exploring various possibilities. But nevertheless every physicist agreed the experiment had to have made an error. *There was no way a particle could travel faster than light, there simply was too much fencing things in.* The public expressed consternation at this resoluteness, with critics of science deriding physicists as closed-minded dogmatists, finally getting their evident comeuppance.

Eventually, however, plain and direct experimental error was uncovered: a loose fiber cable and a very slightly slow clock that together exactly produced the deviation. The following year, experiments by other labs tested and confirmed that neutrinos do not travel faster than light. Theory was so deeply constrained that there was no other choice.

Even when a wider array of theoretical descriptions are possible, the expanse of underdetermination can still itself be sharply constrained. Once arbitrarily contrived contortions are ruled out, underdetermination ceases to be a magical phenomenon that licenses anything.

Some descriptions simply will not work.

> "Reality, Barnes writes, 'will tolerate alternative descriptions without protest. We may say what we will of it, and it will not disagree.' (When a child dies because prayers to the goddesses of smallpox or AIDS or a myriad such diseases do not work, is this not reality speaking, loudly and clearly?)"
>
> Meera Nanda, *Prophets Facing Backward*, 2004

A final fallback set of arguments around epistemic barriers, admits the presence of knowledge in some domains but retreats to asserting the *particularity* of knowledge, asserting abrupt limits to those domains. This is, again, essentially just another case of throwing out induction:

> Sure, our knowledge of different aspects of reality and the sciences may have moved towards unity, with different arbitrary realms of observation converging with one another at their interfaces, but who's to say that this will continue??

In one common variant of this view, the electromagnetic field is just a contrived conceptual tool popularly adopted by one species of laborer, the physicist, involved in one specific craft, that has no universal relevance or influence. To assert that the concepts of a physicist or the workings of a mathematician could impose hard constraints on the conceptual domains of other workers is just *rude*. To claim knowledge more generally or even universally is a violation of others' intellectual property, their territorial sovereignty.

> "There is no such thing as knowledge—what would it be? There is only know-how. In other words, there are crafts and trades."
>
> Bruno Latour, *The Pasteurization of France*, 1988

Of course there's rarely any sort of argument for why understanding should be innately fragmented forever that doesn't depend, implicitly or explicitly, on the ontological claim of Antirealism4: that there is no universality or generality to reality itself. All that is left beyond rejecting induction entirely is to fall back on moralistic resonances, as with Latour's pretense of privileging salt-of-the-earth craftsmen.

What's notable is that such *reverses* the historical political valences of particularism and universality. As I wrote in *Science As Radicalism*, the dawn of modern physics arose from workers horizontally sharing knowledge across contexts to find commonalities, whereas those fields that operate more in a particularist vein, like lepidopterology, arose from aristocratic collectors of curiosos.[85] The notion that each collector of facts should have supreme authority over their own bounded domain, is a studiously aristocratic and bourgeois sentiment, rarely a working class one.

We will get to questions of political stakes and the alleged moral imperatives against knowledge of reality, but note the way that this refusal to accept we can know reality at all is forced to slide into an extreme anti-radicalism or anti-reductionism, which in turn—if it does not embrace the absurdity of an Antirealism4 claim directly—has to leverage reactionary sentiments against solidarity and mobility across particularized contexts. The push towards generality in our knowledge is thus implicitly *moralistically* rejected.

> "The will to transparency is founded on the desire to remove things from their context. To know a thing in itself is the equivalent of radically decontextualizing it."
>
> Sarah Franklin, *Social Text*, 1996

While this outrage towards transparency is partially grounded in an appeal to the daily experience of academics in subjects where true universals don't exist to be found, evidence is weak, and lossless compression is simply impossible, it also appeals to the still lingering influence in the humanities of figures like Heidegger who were violently at odds with the context-transcending cosmopolitanism of modernity.

To shed off the particularity of one's historical or social context, to move towards some conclusion commonly reachable from many starting points such that origins no longer matter, is simply assumed to be abhorrent. The worker must not strive to leave some parochial context, nor raise their eyes above it, but instead be firmly locked back in it.

---

85 William Gillis, "Science as Radicalism," *Human Iterations,* August 18, 2015, https://humaniterations.net/2015/08/18/science-as-radicalism/.

ANTIREALISM6: MODELS OF REALITY ARE ONLY CONTINGENT UPON SOCIAL RELATIONS

Antirealism6 is much like Antirealism5 but emphasizes the *social* contingency of our models, to the point where nothing else gets through. Historical context is said to so completely condition everything that no eternal or universal truths whatsoever can be grasped towards. Social and institutional power dynamics are so complete as to make all our ideas entirely a product of them alone. And the paradigms of different cultures or societies are so fundamentally alien and irreconcilable that they cannot speak to one another meaningfully, cannot converge upon common models, and radically disagree on the most basic epistemic values.

The natural conclusion is that our best models of the world not only capture no degree of "truth" regarding it but are not even influenced by any structures and regularities the physical world might have.

> "Objects of discourse do not exist. The entities discourse refers to are constituted in it and by it."
> Barry Hindess and Paul Hirst,
> *Mode of Production and Social Formation*, 1977

> "The natural world has a small or non-existent role in the construction of scientific knowledge."
> Harry Collins, *Stages in the Empirical Program of Relativism*, 1981

> "It is not the regularity of the world that imposes itself on our senses but the regularity of our institutionalized beliefs that imposes itself on the world"
> Harry Collins, *Changing Order*, 1985

> "As we come to recognize the conventional and artifactual status of our forms of knowing, we put ourselves in a position to realize that it is ourselves and not reality that is responsible for what we know."
> Steven Shapin and Simon Schaffer,
> *Leviathan and the Air-Pump*, 1985

Typically the approach is this: first prove the utterly trivial and never-once-contested point that social context can influence the processes of theory construction and confirmation; then take that to mean that the regularities uncovered by physics are entirely downstream of social considerations and thus entirely determined by sweeping historical or cultural context.

Infamously one of the editors of *Social Text*, Stanley Fish, in a haughty response to Sokal's hoax, compared the facts of physics to the rules of baseball.[86] Both are formed by processes of social construction but thereupon become "*real*" in the sense of fixed. Both baseball and physics models involve some level of social interaction in our convergence upon them, so *who's to say that our understanding of the hubble expansion of the universe is any less arbitrary than our agreement on how many innings we will play.*

> "We now know that our convictions about truth and factuality have not been imposed on us by the world, or imprinted in our brains, but are derived from the practices of ideologically motivated communities"
> Stanley Fish, *Consequences*, 1985

Antirealism6 takes the social background as *primary* in a significant change from the more abstract epistemological issues tackled by Antirealism5. It functionally asserts the total supremacy of the social over the physical and, along the way, strips all intellectual autonomy from the individual.

> "All knowledge is a form of social relations and is discursively constituted."
> Stanley Aronowitz, *Science As Power*, 1988

Of course anyone who pauses for a second will realize that lions and infants make maps of reality and thus have pre-discursive knowledge, but the rhetorical move of subsuming everything under the social allows the antirealist to leverage the chasm many feel between their ways of thinking and that of other cultures.

In his book, *The Order Of Things*, Foucault starts with a now infamous reference to *The Analytical Language of John Wilkins* wherein Jorge Luis Borges invented

> "a 'certain Chinese encyclopaedia' in which it is written that 'animals are divided into: (a) belonging to the Emperor, (b) embalmed, (c) tame, (d) sucking pigs, (e) sirens, (f) fabulous, (g) stray dogs, (h) included in the present classification, (i) frenzied, (j) innumerable, (k) drawn with a very fine camelhair brush, (l) et cetera, (m) having just broken the water pitcher, (n) that from a long way off look like flies'"
> Michel Foucault, *The Order Of Things*, 1966

This parade of grotesque orientalism is all to instill in us a perverse fascination with how alien of creatures these ancient Chinese must have been, how utterly divorced their ways of thinking from our own, and—importantly—how unmoved they would be by our pressure towards coherence, unity, and

---

86 Stanley Fish, "Professor Sokal's bad joke," *The New York Times*, May 21, 1996.

simplicity in the metastructure or ordering of concepts. Borges uses this invention to assert that "*there is no description of the universe that isn't arbitrary and conjectural,*" and Foucault is more than happy to take up the baton:

> "There would appear to be, then, at the other extremity of the earth we inhabit, a culture entirely devoted to the ordering of space, but one that does not distribute the multiplicity of existing things into any of the categories that make it possible for us to name, speak, and think."
>
> Michel Foucault, The Order Of Things, 1966

Such oriental creatures, we are told, slice reality apart, but at different "joints," haphazardly and arbitrarily tied to practical daily considerations of a specific culture. Such an off-putting taxonomy can thus be used to suggest our own models of the regularities of nature reflect little objective structure, reflecting similarly arbitrary practical uses.

Contemporary Chinese philosophers and theorists have been—as you might imagine—deeply critical of this entire passage as well as what Foucault makes of it. And it should likewise be no surprise that the account of this encyclopedia that Foucault engages with, as if it proves anything, is trivially untrue[87] (the same goes for many historical examples central in his works like the "ships of fools" in *Madness And Civilization*). But such base concerns as mere facticity or the voices of foreigners are insulting constraints to many academics, so Foucault's parable and its moral lives on.

Of course, Foucault was not as staunchly antirealist as many of the postmodernists who would take inspiration from him; he largely avoided confronting physics and mathematics, confining his substantive critiques to less radical, *merely empirical*, studies like economics, psychology and biology that are obviously shaped to varying degrees by human-centric practical considerations. But his approach—in line with many other French theorists like Bachelard—was to divide up human history into relatively discrete epochs wherein wildly different conceptual frameworks dominated.

> "In any given culture and at any given moment, there is always only one episteme that defines the conditions of possibility of all knowledge, whether expressed in a theory or silently invested in a practice."
>
> Michel Foucault, The Order Of Things, 1966

For folks in the sixties hoping for a drastic revolution in culture and thought, this sort of historical narrative held immense attraction. Just before Foucault's *The Order Of Things*, Thomas Kuhn had blown up with a similar

---

87 Roel Sterckx, "Animal Classification In Ancient China," *East Asian Science, Technology, and Medicine* 23 (2005): 26-53.

insistence upon unbridgeable differences between eras of scientific thought or theory. These *paradigms* operated very similarly to Foucault's *epistemes*, and Kuhn's watchword was *incommensurability*, a supposed total inability to translate from one paradigm to another, leaving them intrinsically mutually unintelligible. A switch from one to the other could only function through a leap of faith.

And this was very quickly reduced in turn to social context,

> "Like the choice between competing political institutions, that between competing paradigms proves to be a choice between incompatible modes of community life."
> Thomas Kuhn, *The Structure of Scientific Revolutions*, 1970

Yet, just as Foucault shied away from the extreme antirealism that his audience desired—decrying the French intellectual scene and buddying up to analytic philosophers before his death in the US— Kuhn was likewise perturbed by runaway antirealism and retreated from true incommensurability, emphasizing that all societies share five trans-paradigmatic epistemic values.

While Kuhn and Foucault were both focused on periodizing European and scientific history, their appeal to epistemes and paradigms implied and encouraged thinking in terms of a spatial separation of cultures. And the violent stripping out of Europe's deep entanglement with science in India, China and the Islamic World in their historical accounts significantly paved the way.

As every commentator on this notes, such claims of incommensurate, unbridgeable particularism between epochs and cultures had only a couple decades prior found their apex expression with the nazis:

> "There is a German and a Jewish mathematics, two worlds, separated by an unbridgeable chasm... There is no self-sufficient mathematical domain that is independent of ideology and life."
> Ludwig Bieberbach, *Personality Structure And Mathematical Creation*, 1934 [88]

> "There is not and cannot be number as such. There are several number-worlds because there are several cultures"
> Oswald Spengler, *The Decline Of The West*, 1918

> "In the historian's view there is only a history of physics. All its systems now appear to him as neither correct nor incorrect, but as historically, psychologically conditioned by the character of the epoch, and representing that

---

[88] Abraham A. Fraenkel, "How German Mathematicians Dealt With the Rise of Nazism" *Tablet*, February 8th, 2017, https://www.tabletmag.com/sections/arts-letters/articles/hitlers-math.

character more or less completely."
Oswald Spengler, *The Decline Of The West*, 1918

Unfortunately, just as so many impulses like Heidegger's embrace of *thrownness* were laundered into leftwing humanities, struggles for decolonization were systematically subverted by the infestation of nationalism.

One the main impacts of imperialism was to slice apart irreducibly intermeshed global society into discrete groupings that could be more easily managed. This is the origin of nations and borders, an endless horrorshow aptly detailed at great length by anarchists like Nandita Sharma.[89] The social construction of "nations" out of the infinitely more complex and connected social network of individual-to-individual relationships necessarily involved immense violence.

And leftists of the '60s—like the maoists of *Tel Quel* in Paris—were caught up in this bloodlust. Talk of incommensurate epochs of thought easily fit with notions of incommensurate national worldviews. Thus could Peter Lamborn Wilson lavish praise on the authoritarian Shah of Iran for defending a distinct Persian cultural spirit or episteme and then, in turn, Foucault praise the authoritarian Mullahs who deposed him on the same grounds.

The continual slaughter and repression of individual leftists and anarchists on the ground simply didn't register when the true struggle was between supposedly distinct cultural worldviews. Any expression of transnational solidarity by those on the bottom was thus treason to their cultures, and an expression of imperialism.

Likewise, to assert something as more truthful or even to assert some epistemic strategies or values as more productive towards truth was framed as racist and imperialist.

> "Since the relations of science, magic, and religion are internal to each other because they all purport to offer adequate explanations for natural and social phenomena, it is rank ethnocentrism to claim that one may be privileged over the others."
>
> Stanley Aronowitz, *Science As Power*, 1988

> "Modern science is nothing more and nothing less than western science, a special category of ethno-science. In fact, it's too readily assumed universalism has had disastrous consequences for other ethno-sciences."
>
> Claude Alvares, *Science, Development and Violence: The Revolt Against Modernity*, 1992

---

89 Nandita Sharma, *Home Rule: National Sovereignty and the Separation of Natives and Migrants* (2020).

This lens is the epistemic version of the ontological patching of Antirealism4: there might be one underlying universe we're all interacting with, but there are different epistemic approaches and criteria in different cultures with no way to sort out which are better. One culture might consider a belief valid if it helps you predict how the actual universe behaves, another culture might consider a belief valid if it helps keep the social hierarchy intact. Who is to say what really counts as "rationality" or optimal epistemology? What's to compare and contrast the effectiveness of epistemic strategies?

> "...all justifications stop at some principle or alleged matter of fact that has only local credibility."
>
> Barry Barnes and David Bloor, *Relativism, Rationalism and the Sociology of Knowledge*, 1982

> "Lacan's discourse theory suggests that there are as many different claims to rationality as there are different discourses."
>
> Bruce Fink, *The Lacanian Subject*, 1995

The relativist argument here is that you allegedly need a background system of epistemic norms in order to *justify* a particular system of epistemic norms, since this is circular, surely we must then conclude that choice between systems is impossible except *via* parochial attachment or leaps of faith. And, indeed, when two entirely closed self-justifying systems disagree there may be no meta-system to appeal to that could conceivably prove to one that the other is better, since any disagreement with meta-systems introduces an infinite regress of justification, leaving each of us isolated and treading in the void.

But is this actually the situation we find ourselves in?

It's easy to sympathize with certain relativists as merely individuals ground down by despair over the ineffectiveness of discourse and debate,

> "There may be no shared, fundamental value that is binding enough to serve as common ground, in which case there is no basis for discussion. For me this came up recently in conversation with a Christian Scientist about faith healing. It rapidly became clear that our ideas of what would constitute valid evidence for the efficacy of prayer were so far apart that further talk was useless. There seemed to be no underlying concept or value we had in common that could have resolved the difference."
>
> Thomas Clark, *Relativism and the Limits of Rationality*, 1992

And yet, Christian Scientists, despite residing in a wildly different paradigm, can and do abandon their disbelief in the physical world. Indeed from once having a membership of millions, Christian Science has been hemorrhaging members, in particular the younger generations born into the norms of the cult.

Is this an irrational leap, a jump between entirely separate epistemic frameworks, caused by nothing but sociological factors like wider social trends or maybe the incentive structures of the cult's governance? Or can we say that there is an appreciable friction that bad epistemics inevitably suffer over long-term engagement with the fixed regularities of physical reality?

Are not all minds, by virtue of existing in the same physical universe under the same physical constraints, ultimately constrained in their epistemics?

There is of course substantial variation in people's beliefs across our planet. But difference in beliefs, even differences in strategies pragmatically adopted to reach truth, do not imply absolute differences in our emergent epistemic values. The particularity of our experiences, the path-dependency of our experimentation with different epistemic strategies, the limits of our vantage points, all mean that we all operate from different sets of evidence. When two individuals confront ostensibly the same situation they are not actually evaluating the same evidence. Different personal experiences, different currents of historical testimony, different positions within a social graph of trust, will spit out different conclusions. The existence of such differences—even very extreme ones—is inherent, it's *the emergence of commonality* that warrants attention.

Bundles of epistemic strategies are not totalizing and fixed. Just as there are no sharp divisions in level that can be made between epistemic values, strategies, and conclusions; so too is self-reference in justification not closed in a loop, it operates in spirals across meshes of potential epistemic strategies. At his best, Feyerabend recognized this. Our grasping for better *models* of the world cannot be codified in a single easily codified formulism when it comes to *epistemic strategy*. Of course this is a triviality to any scientist: in different contexts there will be different epistemic tools that prove themselves appropriate. You can't do "direct" experimentation on distant stars, you build networks of empirical evidence and theoretical entailment in other ways. But this is very much not to say that "*anything goes*," since any breadth and depth of engagement with reality forces not just a whittling down of models but of epistemic strategies or tools.

The most common historical example leveraged by antirealists has been Galileo's argument for heliocentrism and his resulting conflict with Cardinal Bellarmine.

In Feyerabend's telling (and echoed by many others like Rorty), Galileo "counter-inductively" rejected established theory and clear experimental disproof of heliocentrism because he had irrationally leaped into an entirely different paradigm, alien and unthinkable to Aristotelians. To contest the motion of planets, Galileo had to *also* contest the Aristotelian theory of motion and whether new-fangled telescopes worked, plus whether their observations should count. These two models are alleged to be incommensurable, without a common way to measure them against one another. Since Galileo could not force a complete transition from the tightly-interconnected Aristotelian framework to a still unfinished alternative, he was thus confined to communicative chicanery to confuse Aristotelians that something ultimately totally unjustifiable in their system was in fact entailed by it. Bellarmine—from his just as valid epistemic vantagepoint wherein the bible constituted incredibly strong evidence—correctly rejected these appeals. And there was no common higher epistemic meta-system both agreed to that could render a judgment.

It's important to note that much of Feyerabend's characterization of Galileo involves opportunistically slicing out much of his own words, like a core argument regarding resolving open questions of Mars and Venus, to make it appear as though there was a much less compelling case for heliocentrism.[90]

But it is also a trivial fact that Galileo and Bellarmine were *not* each operating from complete, closed, and fundamentally different epistemic frameworks, they demonstrably shared a vast amount of values, strategies, and theoretical conclusions. They both believed in object permanence, for example, a pretty big and deep theoretical model to hold in common, and Bellarmine did not actually believe that the heavens operated by different principles to which vision didn't extend.[91] Some differences in weightings around pragmatic strategies were not easily transcended through the limited and politically fraught arena of human discourse, but this does not mean that either person's epistemic bundles were truly incommensurate, separate, closed, and fixed.

Even someone assigning very high trust in the inerrant word of divine revelation is applying an epistemic strategy that has developed in light of a given upbringing, set of experiences, social pressures, etc. But that context can break down or be wandered away from. Indeed it takes *great effort* to sustain the sort of context where an epistemic strategy like full deference to authority or disbelief in matter is optimal. Certain commonalities are pressed into all active minds by the fact of our common universe, insofar as generality in that universe becomes relevant.

Galileo and Bellarmine shared a great deal of epistemic commonality. The conflict between them pressed *deep* into networks of differing epistemic strategies, models, and available evidence, but this merely increased the complexity

---

90 Peter K. Machamer, "Feyerabend and Galileo: The interaction of theories, and the reinterpretation of experience," *Studies in History and Philosophy of Science Part A* 4, no. 1 (1973):1-46.

91 Paul Boghosian, *Fear of Knowledge: Against Relativism and Constructivism*, (2006),104

of the *translation* between their two perspectives, it did not create fundamental barriers to it. Something being *practically* difficult to evaluate amidst given particulars and within a single human lifetime does not mean it is beyond evaluation *in principle*.

Feyerabend came to despise the relativist sociologists, to say nothing of the implicit assumptions that "*cultures are well defined and strictly separated*" driving many of their relativistic arguments.

> "As a matter of fact I would say, exactly as Kuhn does, that 'the claims of the strong programme' are 'absurd: an example of deconstruction gone mad.'"
> Paul Feyerabend, postscript to
> 3rd edition of *Against Method*, 1993

But he derived popularity from the same base as them, and so he was obliged to contort to endorse their language, eventually coming back and reconciling himself to the term "relativism."

> "That sounds like the strong programme of the sociology of science except that sculptors are restricted by the properties of the material they use. Similarly individuals, professional groups, cultures can create a wide variety of surroundings, or 'realities'—but not all approaches succeed: some cultures thrive, others linger for a while and then decay. Even an 'objective' enterprise like science which apparently reveals Nature As She Is In Herself intervenes, eliminates, enlarges, produces and codifies the results in a severely standardized way — but again there is no guarantee that the results will congeal into a unified world. Thus all we apprehend when experimenting, or interfering in less systematic ways, or simply living as part of a well-developed culture is how what surrounds us responds to our actions (thoughts, observations, etc.); we do not apprehend these surroundings themselves."
> Paul Feyerabend, postscript to
> 3rd edition of *Against Method*, 1993

In admitting reality has a whittling effect on epistemologies, he was still loath to admit the whittling can be pointed to in terms of evidence and pressures of evaluation, or that it can bring certain conclusions to *any* commonality, gesturing at Antirealism4. All he has, however, is simply the protestation that the noose of reality has only closed *so far* on the space of models and epistemological strategies; who is to say it won't tighten any further?

While Feyerabend, like Kuhn, recoiled from following the relativist sociologists in so extremely playing to the antirealist masses, he shared the same deflection as them. A consistent strategy of Barnes, Bloor and Henry was to point out that there is some in-eliminable infinitesimal degree of uncertainty, we can never be 100.0000% certain about anything,[92] there are always some dynamics of friction, some fraction of concerns still open.

The militant relativist Steve Fuller engaged in even more extreme contortions. To deny the emergence of objective knowledge, he acknowledges that physicists are relatively convergent on much about the universe. But he then asserts that since *everyone knows* that acrimony and disagreement is the marker of a healthy discursive field and epistemic convergence can only emerge by social tyranny, self-censorship, etc., this very convergence *invalidates* the field of physics, negating any claims it could make about the universe. Having written off the entirety of physics as proven to be compromised by its agreement, one can then point to any disagreement and messiness in less radical fields as a challenge to realism.

Yet as the relativists doubled-down, they were forced to proclaim that even evaluative criteria like simplicity or consistency were merely cultural and regional. Even the perceived regularity *of day and night* arises only from social convention; change the culture and no one would perceive any cyclicality to it. Beyond trying to write off simplicity as an objective epistemic virtue, Barnes, Bloor and Henry wanted to treat *induction itself* as an exclusively social phenomenon, impossible in isolation for an individual mind. They followed in the footsteps of Collins by claiming they merely denied a *"logical compulsion, not necessarily psychological and/or sociological compulsion."* Thus picking out regularities and variances in compressive success is slid into the realm of *"psychological"* without much justification.

It's important to note that while the relativists made broad declarations that our models of the universe were shaped exclusively by social factors rather than the physical world and leveraged claims of supposedly deep and unbridgeable differences between cultural epistemologies, many retreated from this in various ways.

The Strong Programme were prone to some of the strongest philosophical claims but at the same time they would fall back on the whine that they were merely playing with one specific methodology—a relativistic one—to see what fell out. Their infamous "symmetry" approach argued that the sociologist should evaluate how a given belief about the world was arrived at without any reference to whether the belief was actually true.

Now, it's important to note that there is a limited sense in which such an approach is justified, one should be able to explain how both Mendel and Lysenko reached their conclusions in the same manner, without lazily retro-

---

92 We simply can never have sufficient evidence for justified true belief in such infinite certainty or credence levels. A zealot can wall off a *false* "belief" strongly, but cognition is a physical process in a system that is never infinitely rigid.

actively telling a just-so story based on one side's results.

The problem is that *sociological* explanations for why a given group of people believe a thing must appeal only to *social* causes and make no direct appeal to the actual material facts. Thus the "symmetry" principle is really an asymmetry: a declaration that sociology *on its own* can provide a complete description of the origins of a belief without recourse to another realm of investigation involving pre-social physical causes and evidence.

> "If the world must be introduced then it should play no more role than the fire in which pictures are seen."
>
> Harry Collins, *Changing Order*, 1992

This drew outrage from the science warriors.

> "If the claim were merely that we should use the same principles of sociology and psychology to explain the causation of all beliefs irrespective of whether we evaluate them as true or false, rational or irrational, then I would have no particular objection (though one might have qualms about the hyper-scientistic attitude that human beliefs are always to be explained causally through social science). But if the claim is that only social causes can enter into such an explanation—that the way the world is cannot enter—then I cannot disagree more strenuously."
>
> Alan Sokal, *What the Social Text Affair Does and Does Not Prove*, 1997

Sokal devastatingly proposed a sociological experiment to resolve the question: we simply erase some knowledge and re-run societies to see if the inverse square relation in gravity and electromagnetism is re-discovered in radically different social contexts (instead of an inverse cube relation).

Even in the infamous issue of *Social Text* where Aronowitz desperately tried to gather supporters, the feminist Hilary Rose chimed in with standard leftist points but refused to take the party line, calling the approaches of the Sociology of Scientific Knowledge camp "clumsy," condemning its reduction of science to "myth," and admitting it frequently slips from epistemic relativism to ontological relativism.

But it also drew fire from within the ranks of the sociologists.

> "Are objects allowed to make a difference in our thinking about them? The answer given by David and repeated over and over again by all the descendents of this tradition—even the empirically minded ones such as Shapin, Schaffer and Collins—is a resounding no."
>
> Bruno Latour, *For David Bloor... and Beyond: A Reply to David Bloor's 'Anti-Latour'*, 1999

Can we fully explain climate scientists' beliefs about global warming without making *any* reference to material evidence? Do climate deniers really engage with material evidence to the *same breadth and depth*?

As time went on, a number of STS folks have broken with the cult of Bloor, with individuals like Erik Baker and Naomi Oreskes criticizing the "*science as a game*" framework and writing the prominent STS book *Merchants Of Doubt* in a very sharp violation of the Bloorian symmetry principle. Similarly, autopoietic systems theory (AST), in contrast to SSK, argues that science (as a distributed social dynamic) is a recursive system particularly close to the material environmental conditions and substrates.

Since many students come to STS now with climate denial as a central concern, there's been a widespread rewriting of this history in modern classes, with "*relativism*" often redefined in the most anodyne senses and pretty much all content or context of the science wars erased.

Yet the holdouts have not been happy to see their colleagues retreat. From the start, Steve Woolgar chastised other relativist sociologists for their hesitancy in attacking realism directly, specifically that *"the world exists independently of, and prior to, knowledge produced about it,"* arguing instead that objects are not the origin of conceptual representations, but conceptual representations are the source of objects,

> "Despite the apparent radicalism of its stance on the relativity of scientific truth, this ambivalence raises the possibility that SSK has done little to revise basic ontological commitments. Indeed, recent work in SSK has been dubbed 'epistemologically relativist and ontologically realist.' This is curious given that a major thrust of post-modern critiques of science is to suggest the essential equivalence of ontology and epistemology: how we know is what exists... Our ability to speak as if realities exist independent of our knowing them is a key function of language and representation. But can an object exist independent of our practices of representation?"
>
> Steve Woolgar, *Science: The Very Idea*, 1988

Finally there's what most antirealism boils down to behind the scenes: there may be some structure, and some people may perceive it to some degree, even through science, but *it would be better if they didn't* and we had better not lend any credence to this very dangerous or disruptive activity. We should even go so far as to shift our language so that *direct reference* to reality is as forbidden or torturous as possible.

ANTIREALISM7: KNOWING REALITY IS IMMORAL

Antirealism7 argues that knowing itself is bad. This is a very widespread perspective, common far beyond postmodernism, and treated by many in the humanities as trivially proven following the Manhattan Project. In Jerome Ravetz's words, the nuclear bomb left physics' quest for truth *"tarnished beyond repair; and through its status as the paradigm field, its moral fall affected the rest of science as well."* [93] Even many militant opponents of postmodernism like anarcho-primitivists, who saw in postmodernism an abandonment of concern with raw material reality for language fetishism, at the same time thought that breadth of knowledge of the natural world was a bad thing that inherently unleashes ecocide.

It's important to distinguish this from critiques of "scientific knowing" that engage in crude or absurdly contrived gerrymandering of "science"—often in collaboration with the bloodsoaked rulers seeking to appropriate its mantle. Take for example, Vandana Shiva confidently declaring that *"eighty percent of all scientific research is devoted to the war industry and is frankly aimed at large scale violence."* [94] Such a summary is beyond ludicrous—conflating crude engineering or technological development with the radical inquiry of actual science, and presumably quantifying the amount of "scientific research" performed in terms of something arbitrary like funding. One could easily counter that only fundamental physics counts as science or that the continued efforts of every infant learning object permanence from scratch constitute a much vaster (if redundant) expanse of net scientific research. But when such critics of science agree with fascists, corporatists, or militarists in bespoke redefinitions of "science," they are not directly asserting that all knowledge of reality is bad, merely the subset they allow to be called "scientific."

Yet, the wider sentiment that "man was not meant to know reality" still crops up across diverse ideologies and cultural spaces as a blanket prohibition or warning. This is usually grounded in concern with the consequences of being able to move matter around in more ways and with greater effectiveness.

But, of course, if knowing even our arms and fists is bad because such knowledge would enable us to punch other people, then the person asserting this has implicitly conceded that reality exists and can be sometimes known. This moral sort of antirealism is often trotted out to denounce realists, and often in the same breath that someone asserts some other variants of antirealism, but, while ever present as a background cultural assumption, is often less

---

93 Jerome Ravetz, *Scientific Knowledge and its Social Problems* (1971), 60.
94 Vandana Shiva, "Reductionist science as epistemological violence", in, *Science, hegemony and violence: a requiem for modernity*, ed. Ashis Nandy (1988).

centered, since the contradictions with other antirealisms are apparent to most.

> "Whatever heights science may attain, it may only make more and more patent two facts: 1. Those heights can only be attained by mercilessly crushing and walking over mountains of human beings. 2. And indeed be it for this single reason, all of planet Earth should become as quickly as possible a radioactive desert or disappear through disintegration."
>
> Louis Wolfson, *Full Stop for an Infernal Planet or The Schizophrenic Sensorial Epileptic and Foreign Languages*, Semiotex(e), 1978

ANTIREALISM8: LANGUAGE SHOULD NOT REFER TO REALITY

> "I once attended a conference of literary critics at which one speaker was talking about Jean-Paul Sartre's account of his own childhood. A deconstructionist asked her, in a pained and patronizing tone, whether she was claiming that there really had existed such a person as Jean-Paul Sartre, independently of what we might say of him. When she said yes, she was, she at once lost the attention of the deconstructionist contingent."
>
> Andrew Collier, *Critical Realism*, 1994

Antirealism8, in contrast to Antirealism7, worries not about knowledge but about permitting the very notion of reality to circulate within society. This moral objection runs throughout postmodernism but was put into sharpest relief by Rorty in his *Pragmatism As Antiauthoritarianism*, where he compares believing in reality to believing in God and says it's an innately authoritarian perspective. If you think that one claim is True and another claim is False, you are, after all, creating a hierarchy between them. If you assert that there's only one reality, you've created a spiritual centralization, a monopoly just like The State! (Rorty, being a standard pluralist liberal, is firmly in favor of the actual existing state, with its actually existing police and prisons, but never you mind that, you see he's an anti-authoritarian where it counts: in poetic abstractions that can circulate as edge points in academic towers.)

The way this often functions in practice is a constant stepping back from directness in linguistic reference. Like, for example, refusing to engage with or acknowledge ontological concepts and discourse, and always aggressively redefining such in an epistemic frame. Whenever a term refers too obviously or innately to a physical regularity, a set of folks desperately step in and work overtime to push that term back into the structuralist net of social and observer contingent constructions. "World" is corroded from meaning, you know, the singular physical universe we're all a part of, into an actor's conceptual baggage, their lenses, preconceptions, and operating assumptions. Often this pivot in definitions will happen in a single sentence:

> "Though the world does not change with a change in paradigm, the scientist afterwards works in a different world."
>
> Thomas Kuhn, *The Structure Of Scientific Revolutions*, 1970

I keep reaching for horribly mangled phrases and neologisms like "*physical regularities*" in this book because the fuckers have spent decades colonizing and redefining every single term we might have to refer to such, so that "*reality*" and "*truth*" are likewise whittled away in common use from any sort grasp onto anything outside the social (or epistemic).

"I believe it's *your* truth that you were raped; stop trying to make it *my* truth."
Ancient leftist proverb

You think, oh there's little harm in letting folks speak in terms of discursive worlds, that certainly has some substantive meaning. But then very rapidly they start treating "world" solely as such. You can't even cede them virtually all the territory and retreat back solely to the last and most unassailable term "reality." Because now even that has been nearly completely conquered, with it becoming hegemonic to speak of "consensus reality," implying that if someone believes in microchip vaccines or that the Holocaust didn't happen, our only critique is not that they are objectively false, but that they are mere dissidents to a social consensus, living in merely a non-standard reality.

You are not allowed to assert that you're telling "the truth" or someone else is "lying" because to do so implies that there can even be a truth. A similar tactic is in the insistence that literally everything is "invention" rather than "discovery," trying to aggressively taboo the use of the latter with regard to regularities and structures in the physical universe. Electrons and entropy are invented, not discovered. Nothing is actually discovered. We make reality!

A.J. Ayer infamously fought with Bataille and Merleau-Ponty in a Paris bar in 1951 at 2am over their confident denial that the sun had existed before humans.[95] Similarly, to Sokal's ire, Bruno Latour loved smirkingly insisting that Ramses could not have died of tuberculosis because the concept of tuberculosis was "invented" after his death.[96] Myriad permutations of this move—from Heidegger's dismissal that Newton's insights were neither true nor false before he was born to Žižek's declaration that we should not directly say that dinosaurs existed in the past, but only that we currently believe they existed in the past[97] or Rorty's claim that there is "*no sense*" in which a dinosaur or anything else is "*out there*" before we describe it[98]—permeate through the history of fights between realists and postmodern antirealists.

At the most brute level, this conflict is between whether "the sun" or "tuberculosis" or "dinosaurs" refers to a thing, a physical structure or pattern, or exclusively just a socially-circulating concept. The only appropriate response to that sort of inane move is to roll one's eyes with unbridled disdain, but we

---

[95] Andreas Vrahimis, "Was There a Sun Before Men Existed?" A.J. Ayer and French Philosophy in the Fifties", *Journal for the History of Analytic Philosophy* 1, no 9 (2012) 1-25.

[96] Bruno Latour, "On the Partial Existence of Existing and Non-existing Objects" in *Biographies of Scientific Objects*, ed. Lorraine Daston (1996).

[97] Slavoj Žižek, *Less than nothing: Hegel and the shadow of dialectical materialism* (2012), 647.

[98] Richard Rorty, *Truth and Progress: Volume 3: Philosophical Papers* (1998), 87.

must pretend to take it seriously for a second. The antirealist here is certainly dancing up close to Antirealism3—likely intentionally—but their attack is to collapse the signifier "the sun" to mean the signifier "the concept of 'the sun'" even though those two phrases are very distinct in actual language and mean very different things to almost everyone. It's also, of course, inconsistent, since "before humans" is taken to mean the object-level reality of actually "before humans" and not "the concept of 'before humans.'" The postmodernist has willfully cross-contaminated between two different language games to two consequences: 1) to militantly demand an implicit structuralist presupposition that there is no such thing as referents to physical reality in our language use, and 2) to, at the object-level normal human language interpretation, state that we don't just construct our conceptual scaffolding involved in our understanding of "the sun" but dream the sun itself into existence.

> "It is encouraging a fallacious jump from 'astronomy is constructed' to 'therefor the things astronomy is about are constructed'—or something like that. And when speaking with philosophers Kuhn would deny holding this outrageous belief. But when speaking to historians, he'd go back to using this language."
>
> John Burgess, interviewed in *The Ashtray* by Errol Morris, 2018

It's easy to see how this sort of linguistic maneuvering rhetorically facilitates ontological and epistemic antirealisms. Take this passage from Latour, where he equates "nature" into "nature's representation" in order to underhandedly leverage a banality into an unjustified edict forbidding saying that the actual material facts might at all influence which models win out:

> "Since the settlement of a controversy is the cause of Nature's representation, not the consequence, we can never use the outcome—Nature—to explain how and why a controversy has been settled."
>
> Bruno Latour, *Science in Action*, 1987

Infamously, the postmodernist movement was so widely prone to such conflations in motte-and-bailey fallacies that it provoked the naming and definition of that fallacy by Nicholas Shackel.[99] In the examples he focused on, from Foucault to Bloor, as well as in the generalized examples now usually given as the fallacy entered widespread awareness, the subject is almost always antirealism of one kind or another. And plenty of other philosophers have pointed to similar examples of flagrant "*switcheroos*" across the postmodernist and relativist sociologist literature.[100] I think this is illuminating. It shows not only that we have a staggering

---

99 Nicholas Shackel, "The Vacuity of Postmodern Methodology" *Metaphilosophy* 36 (April 2005): 295-320.
100 André Kukla, *Social Constructivism and the Philosophy of Science* (2000)

paucity of terminology to deal with really common—even populist—antirealist positions, and so slippage between them is easy, but also that slippage is frequently motivated.

When someone like Peter Lamborn Wilson waffled around between different—incompatible!—versions of antirealism, he surely experienced this as a righteous kind of play. Demanding that he be consistent or coherent is un-fun (which is, in turn, a grave violation of implicit moral imperatives). Almost identical motte-and-bailey moves still flourish at the heart of chaos magick and other occult brainrot popular in subcultural spaces—often borrowing the exact same slippages between various antirealisms. It is simply wildly insufficient for critics of postmodernism to point out that their targets are performing disingenuous motte-and-baileys, as so many science warriors in the Science Wars did, without addressing the motivation behind them and establishing why someone should more strongly feel different motivations.

It's worth going into more detail about the stakes.

5

THE STAKES OF REALISM AND ANTIREALISM

"Reason and power are one and the same"
    Jean-François Lyotard, *Dérive à partir de Marx et Freud*, 1973

"The theoretical enterprise of the west is *imperialistic* in a way! It's got to find a belief that others then have to believe!"
    Rick Roderick, *Nietzsche on Truth and Lie*, 1991

"the notion of universal objective 'truth' or 'reality' is socially constructed and historically tied to 'Western' imperialism and epistemological genocide anti-imperialism necessitates opposing hegomonic 'Western' epistemology and defending marginalized ways-of-knowing"
    an emblematic Fediverse post, 2023

As far as academia was concerned, the Science Wars ended with a long slow decline. Once fiery conferences and editorials petered off into a handful of remaining academics on either side politely debating technical minutia. The sociological relativists retreated to less and less impressive claims, often little more than semantic preference over what our language should bring to attention, while a small crew of realist philosophers composed more intricate and esoteric responses. Legend says Kusch and Seigel are still publishing rejoinders to one another to this day.

Yet the diminishment of public engagement did not signify any sort of rapprochement between the two sides, any lessening of tensions or furor, but rather something more like mutual secession. Even those last remaining souls in contact across the divide remained passionately motivated.

Naturally both sides teach that they unequivocally won and that any somehow lingering remainder of the other side warrants nothing but contempt if not pity. If a young leftist student of a postmodernist professor says something that sounds realist, the response is often a disbelieving scoff. *To think we can form models of the world that can be true and that modern science is better at this is "positivism," a completely discredited antiquity that was responsible for everything from the conquistadors to atomic slaughter and fell apart in the 1950s. No one clings to such things anymore besides the most uneducated hyper-reactionaries.*

Such derision is conjoined with a constant echoing of postmodernist metanarratives in every paper, until they take on the solidity of something that *everyone knows*.

Unfortunately, young leftists with legitimately righteous concerns are usually not in on the game. When broad narratives repeatedly echoed have incredibly sharp stakes for liberation, the studied nebulosity of postmodernist academics is taken to arise not from a need for plausible deniability but from moral timidity, a liberal unwillingness to leap more entirely to conclusions that radical students count themselves free from. What *"everyone knows"* thus blossoms.

A classic example is the ratcheting claims where *science is nothing more than universalism, which is nothing more than christianity, which is nothing more than a culturally specific disease of Europe from which all bad things in the world can be traced.* (Since they're studiously not taught, it's unthinkable that there could be other universalisms from outside of Europe, much less science, atheism, materialism, rationalism, etc. arising from below with the oppressed.) Old decolonial ecofeminists, pickled in their own current of transphobia hostile to "technoscience," echo to one another that Europeans invented a strict gender binary, imperiously speaking over minorities like the hijras in India to cut them out of any common experience of patriarchy and trans femininity.[100] And this in turn empowers ever more shallow rhetoric claiming that *patriarchy* and *gender* were wholesale invented by Europeans. *Just another disease of the enlightenment rationality that killed the witches.* When forced to drag out arguments or sources for such sweeping claims—often under demands from brown folks in cultures with millennia of struggles against local patriarchy—they will appeal merely to sources that accurately document the very real gender-constructing violence and patriarchal horror continuously inflicted by European imperialism and colonialism. *Are you not emotionally moved by that?! Why would you undermine the importance of that by quibbling about whether patriarchy dates back further?! We're in a war! Anything that's not rhetorically emphasizing the side of the oppressed is working for the oppressor!*

This is all very effective at policing the bounds of thought and discourse within the scattered remains of academic postmodernism precisely because the stakes are so big.

But even if the scoffing adjunct or overzealous student may be echoing these defensive moves in relative ignorance, that does not mean that there are no passionately motivating arguments for antirealism beyond campist rhetorical strategy or defensive academic parochialism. Those that have surveyed the philosophical debate and still dug their feet in do so for a variety of specific moral concerns, and are overwhelmingly explicit about that. With major figures talking about antirealism, relativism, etc. as not a belief but a "*stance*" that first and foremost represents moral commitments rather than any concern with pure epistemology.

> "We no longer have to fight against microbes, but against the misfortunes of reason… Because we have other interests and follow other ways, we find the myth of reason and science unacceptable, intolerable, even immoral."
> Bruno Latour, *The Pasteurization of France*, 1984

---

100 Talia Bhatt, "The Third Sex," *Trans/Rad/Fem*, September 1, 2024 https://taliabhattwrites.substack.com/p/the-third-sex.

Truth As "Arrogance"

Driving pretty much every single critique of realism you will ever encounter is a sharp moralistic objection to the *tone* or *personality* they think realism obliges.

Most commonly, the perspective that we can and should model the world with increasing accuracy is said to be an *arrogant* and *controlling* orientation towards the world, for example, Julia Kristeva called science a stance of *"sadistic control"* and Lyotard labeled all of science as *"arrogant."* [101]

> "[Relativists] oppose intellectual imperialism and value epistemic humility or tolerance."
>
> Martin Kucsh, *Relativism in the Philosophy of Science*, 2021

I will admit from the outset I have always found *"arrogance"* an incredibly weird concept. It's clearly a very significant notion to many people, who act like we should all know what it refers to and treat it as a primordial vice, but in common use "arrogance" seems to always blur two very different dynamics together. The first is epistemic overconfidence. The second is a lack of social deference.

It is, of course, an epistemic vice to get the odds of something wrong, overestimating or underestimating its likelihood. Usually this is very situation-specific, but an individual can develop a systematic bias towards certainty in everything they believe, just as they can develop a systematic bias towards timid uncertainty, refusing to believe the overwhelming evidence in front of their eyes. It's plainly absurd to think that either systematic bias is a clear lesser evil across all contexts. Someone who underestimates their likelihood of correctness will lose bets—even if just in the sense of avoiding good bets—just the same as someone who overestimates their correctness.

As any anarchist knows, our world is drowning in hesitant timidity. Most of the horrors we're surrounded by are underpinned less by arrogance than by indoctrination in *humility* holding people back from revolt. The very first thing the liberal education system seeks to beat out of children is any confidence that might lead to action against injustice. And the anti-arrogance moralizing continues up the establishment from the media to the police.

> "I also think that people confuse anarchy with extreme relativism, or philosophical anti-foundationalism, which always makes me a bit nervous... the most relativistic people I've ever met have been cops. I once spent five hours in an arrest bus with about 40 other people in plastic handcuffs and this one police

---

101 Jean-François Lyotard, *Libidinal Economy* (London: Continuum, 2005), 252.

> officer kept coming into the bus to argue with us, a guy we came to refer to, not very fondly, as Officer Mindfuck. He would always take an extreme moral relativist position and say 'Sure, you think you are driven by a moral imperative higher than the law, but your problem is you think yours is the only possible legitimate point of view.' I've seen that a lot of times since: a cop gives an order that's just completely insane, like surrounding you and then ordering you to disperse, you're stupid enough to try to reason with them, they just say 'Oh so you think you have it all figured out, don't you? You have all the answers.'"
> David Graeber, *Anarchy — In a Manner of Speaking*, 2020

Certainly, when Feyerabend proclaims that, "*it is time to cut [scientists] down to size*," amidst his endorsements of leveraging state violence against uppity scientists, his voice resonates with the glee of a riot cop about to crack activist skull.[102]

It's understandable that those in power constantly harrange us for our "arrogance" and treat it like the worst crime imaginable, but it's less clear why anyone on the bottom would care, much less viciously police one another for any potential indicator of it.

The answer, I think, is about dynamics of social deference. An individual giving an account of objective physical reality is resistant to argument or social pressure in a similar way to a tyrant who does not have to listen to a peasant. Listening to the confidence of a scientist thus *feels* identical to the experience of listening to a king issuing edicts, despite everything else in the situations being different. It's the same *tone*. And such tone is taken not as an epiphenomenon of domination but as constitutive of it.

Much of human society is held together by a dance of give-and-take, where parties painstakingly demonstrate they can be amenable to one another's influence. You assert a need, I comply to show I'm attentive to it, then I assert a need and you comply. We shift around on our demands, let them be watered down through haggling. Through this dance, trust is built. When rank-and-file conservatives recoil in insulted outrage over the firmness of climate scientist claims, they are studiously continuing one of the most basic pro-social instincts humans have: to aggressively punish those who do not demonstrate goodwill towards us.

When the marxists at *Social Text* tried to frame their critique of science as merely one of support for democracy, sure they were covering up sharper antirealism, but in another sense they were being very explicit about their underlying objection. The regularities of the universe are coldly antidemocratic; scientific fact is ultimately not up for a vote. In many cases it's not really even up for *discussion*. When the evidence is overwhelming, when there is no theoretical alternative beyond the most contrived contortions, you either sur-

---

102 Paul Feyerabend, *Against Method* (1975), 304.

render before that or you are obliged to go to war with the entirety of science, rationality, and realism.

To their credit a great many postmodernists were quite explicit about all this.

Bruno Latour, for example, framed his opposition to science, to universalism and reduction in sharply moral terms that tried to steal the mantle of antiauthoritarianism. Science is authoritarian, he emphasized, because it does not engage in *compromise*. He emphasized *"compromise"* as a moral good repeatedly.

But antiauthoritarianism is notoriously *uncompromising*, this is its core difference from spineless liberalism. Freedom is not to be negotiated. We do not go to the bargaining table with oppressors; we simply defeat them. We work to make the ability to rule *impossible* for anyone. We stage attacks upon the infrastructure necessary to exert domination; we catalyze norms of insurgent solidarity against power-seekers; we fight back.

Authoritarians like Engels have long griped that such defensive violence somehow constitutes equivalent domination as that of slaveowners because it forcibly *denies* a would-be slaveowner the option to own people. Anarchists retort that there are clear differences around the total amount of freedom provided to folks and a slave can, in fact, know perfectly well who their slavemaster is and the necessity of killing him, without debating that fact before some democratic town council. This may make the slave arrogant, uppity, or audacious, but those are then terms of praise. May such arrogance spread.

Liberals see a union in struggle with a boss and think instinctively that the only anti-authoritarian option is to make them meet and compromise. Anarchists recognize the only truly antiauthoritarian option is ultimately to refuse to negotiate, and sometimes to shoot the boss dead. *"We don't make demands." "We don't want bread or the bakery, we want everything."* We are eternally the least popular position, a perpetual minority punching up, because we offer would-be rulers no retreat, no alternate permutation of power they can switch to. While all other political or ethical frameworks attempt to preserve or "legitimize" *some* arena of domination, anarchists steadfastly work to obliterate *all* domination.

Yes, in both science and antiauthoritarian struggle, to say nothing of daily life, there are spaces where negotiation and concessions are valorous, but there are also firm boundary conditions of immovable truths. Democracy and compromise are not fundamental virtues. Nor is humility. They certainly have no claim to radical anti-authoritarianism.

The old rhetorical device, common across all authoritarians, wherein *everything* is power and so freedom is impossible and resistance is condemnable as another expression of power, is not unrelated to the deliberate conflation of "control" over matter with control over people. While any average person would see informed choice in the world around us as the very definition of freedom—a matter of options, avenues in which to act—the authoritarian tries to twist language so that *looking at nature* is the same thing as a *state surveillance program*. Because they don't notice the explosive agency of other people, they really see no difference between using a hammer and using a slave. *If you object to the constraint of the slave's agency, why don't you object to someone choosing to move a hammer?! Checkmate, abolitionists!!* Postmodernists like Kristeva inherit this conceptual schema but mix it with the heuristic against arrogant tones, concluding that trying to know reality is the real authoritarianism.

Science, Latour complains in *The Pasteurization of France*, is "*so radical, so total, and so absolute.*" [103]

It should not be surprising then that Latour likewise complained bitterly in the '90s when antifascists in France jumped a nazi.

---

103 Bruno Latour, *The Pasteurization of France,* trans. Alan Sheridan and John Law (1998), 215.

"Humility" Before Nature As Tyranny

There is, however, a far less common but still resonant argument wherein the danger posed by realists is said to be one of humility.

The postmodernist who made the most explicit, forthright, and extensive appeals to liberal democracy in opposition to scientific realism was the philosopher Richard Rorty whose final book, *Pragmatism As Antiauthoritarianism*, is a systematic exploration of the moral stakes he felt were involved across the debate. In many ways, this book was conceived of as the realist version of his project, ruthlessly returning tit for tat.

Rorty is often brought up as a proponent of postmodernism unblemished by the rhetorical underhandedness endemic to other writers, as his origin in analytic circles kept his writing notably earnest and free of obscurantism. In part this is because Rorty was not so much directly influenced by postmodernists—he did not come out of their discursive tradition or social circles—but rather saw them as allies. He praised them, cultivated their base as fans, and defiantly wrapped himself in their labels when they were critiqued by other philosophers. Indicative of this difference in origin, his approach to antirealism doesn't center on deriding scientists as arrogant before laypeople, but as humble before an alien outside force.

Like so many others, Rorty hated the very concept of a singular truth or a singular objective physical reality because that smacked to him of a god or king. He was convinced that to allow the very idea of truth to circulate in human discourse was to empower tyranny.

> "I see the pragmatist's account of truth and more generally their anti-representationalist account of belief as a protest against the idea that humans must humble themselves before the will of god or the intrinsic nature of reality."
> Richard Rorty, lectures in Girona, 1996 [104]

Rorty took umbrage with the statement *"our colleagues are our friends, but truth is more our friend"* and saw this as obviously a violation of antiauthoritarianism, at least when "antiauthoritarianism" is taken to imply a communitarianism. What is social is seen by him as by definition egalitarian, and thus a commitment to accurate models regardless of how much it violates or snubs the social consensus is anti-egalitarian. Rorty was explicit in engaging with questions in the seminars that he meant "authoritarianism" as any authority

---

[104] Richard Rorty, "'Pragmatism as Antiauthoritarianism' Part 1. Richard Rorty's 1996 Girona Lectures, with discussions," Richard Rorty's Ferrater Mora lectures, University of Girona, 1996, *Bob Bradom*, uploaded November 21, 2021, YouTube, 4:18. https://www.youtube.com/watch?v=-k-IUoEAHpg.

other than social bonds and that *"respect for the non-human"* was bad.

Pragmatism As Antiauthoritarianism had its origin as a lecture series given in the '90s, and one of the interactions in the Q&As has stuck with me as deeply revealing. A woman in the audience pressed him on whether he would admit lions and gazelles can have more or less accurate maps of objective reality in various situations in ways that are not merely dynamics of some linguistic community, but engage with the structure of the material world. Rorty's response was to bite the bullet that humans are categorically different in living subjective intentional lives, explicitly privileging us over the rest of nature because we can do *poetry*.[105] No more perfect morality play could be written to demonstrate why green anarchists like Zerzan have had so little tolerance for postmodernism.

Rorty's position may seem in deep conflict with the attack on science as arrogant, but it's just another side of the same coin, an outrage at the intrusion of the base physical into the social.

Collectivists have a tendency to conflate being prostrate before one another, self-abdicating, and *mutually* professing to be no more than worms, as the only possible alternative of hierarchical society, so they see the non-negotiable firmness of reality's physical regularities as an aggression. Some see this as scientists *themselves* as introducing firmness, but others, like Rorty, see the scientists as earnestly encouraging humility before an outsider, an alien who does not reciprocate the same social bonds of mutuality.

> "The need for choice between competing representations can be replaced by tolerance for a plurality of non-competing descriptions which serve different purposes and which are to be evaluated by reference to their utility and fulfilling these purposes rather than by their fit with the objects being described. To ask which of these two accounts of the universe is true [religion or science] may be as pointless as asking is the carpenters or the particle physicists account of tables the true one for neither question needs to be answered if we can figure out a strategy from for keeping the two accounts from getting in each other's way."
>
> Richard Rorty, lectures in Girona, 1996 [106]

Rorty was deeply offended by the existence of fundamental physics since its universalist project implies no such strategy of mutual tolerance is possible. Since everything *besides* physics can be cast in terms of art or technology, he wanted to say that the beating heart of science is a rounding error that should be ignored and then, with its radicalism safely expunged, all the weaker sciences be re-framed as nothing more than pragmatic crafts. Unable to get

---

105 ibid, 6:43:58
106 ibid 26:10.

traction against the realism and radicalism of physics he resorted to calling its account of the world "a divinity" and "a god of power."

> "Once you become polytheistic you are likely to turn away not only from priests but from such priest substitutes as metaphysicians and physicists."
> Richard Rorty, lectures in Girona, 1996 [107]

Ironically, of course, as Nanda has repeatedly pointed out, the polytheistic priests in India demonstrate the opposite. The traditional brahamical caste system is organized precisely in opposition to nature. Contact with the firmness of physical reality is relegated to the lowest castes, while tyrannical power resides in precisely those who are exclusively tied to the social. As a result, Indian philosophy has a long tradition of materialist universalism, atheism, etc. emerging as a philosophy of those on the bottom.

Rorty shares the exact same instincts of oppressors to see such universalism as the dangerous uppity noises of those who should be nothing more than janitors whose only purpose should be to better insulate the social from the physical. Any grand aspiration in science beyond technological production must be beaten back into proper submission. Physicists are allowed to operate as adjuncts to semiconductor production but never speak or think of *"what is actually the case in the world"* except in the most apologetic ways patronizingly dismissed by all serious individuals as toy language we allow the kids.

The fact that this position is an exact carbon copy of the stance pushed by the US military complex as it stomped out the far-left and philosophical field of physics in the 1940s doesn't seem to trouble Rorty. The humanities and the technocrats are more than happy to join forces to crush their mutual enemy in science.

---

107 ibid 1:51:32.

Representation

Just as Rorty decided to appropriate the mantle of anti-authoritarianism for his authoritarian liberalism, so too in 1989, did a post-structuralist philosopher with absolutely no background in or knowledge of the anarchist movement at the time, Todd May, decide to declare himself an "anarchist."

This was the culmination of a rising postmodernist fad of characterizing something they would call "representation" as the ur-evil behind all other societal ails. Examples of this abound from Latour's collaborator Steve Woolgar denouncing all science and reason as stemming from the "*ideology of representation*" [108] to the French professor Daniel Colson eventually attaching it to his reactionary rants about political correctness and the plight of white males in activist circles. May took this faddish language and declared that opposition to representation was the heart of "anarchism."

When this weird fixation on an overly broad metaphor was received coldly by actual anarchists, he responded by denouncing them as insufficiently "anarchist."

> "Post-structuralist theory is indeed anarchist. It is in fact more consistently anarchist than traditional anarchist theory has proven to be. The theoretical wellspring of anarchism—the refusal of representation by political or conceptual means in order to achieve self determination along a variety of registers and at different local levels—finds its underpinnings articulated most accurately by the post-structuralist political theorists."
>
> Todd May, *The Moral Theory of Poststructuralism*, 2004

Putting aside the tortured hypocrisy of denouncing representation while imperiously seizing the representative mantle of anarchism from anarchists,[109] it should be obvious that such a use of "representation" involves a weird equivalence between two very different concepts. In this case, blurring the obvious tyranny of *political* subordination where a politician "represents" their constituents with an individual applying *conceptual* models to "represent" the world.

We can all agree that the politician who claims to speak *for* a group of people inherently does violence to those folks. And one component of this

---

108 Steve Woolgar, *Knowledge and Reflexivity* (1988).
109 A move that springs anew eternally among continental philosophers, mining us for aesthetic prestige and edginess, then turning around and writing introductions to anarchism that are just a couple old dudes they found mixed with their non-anarchist colleagues, aggressively devoid of anyone so declasse as to be involved with the living movement. And then they're surprised when we laugh at them rather than bowing before them. Catherine Malabou is a noteworthy example, in that she even tried to appropriate the opprobrium at her own behavior.

is that, regardless of the intent or earnestness, no politician can access, much less compress or convey, the vast array of preferences and perspectives of their constituents. To "represent" another person in this sense is to cut off their voice and insert your own. Many of the most common power dynamics we deal with everyday involve such representation. The self-appointed "black leadership" that walks with the police commissioner up to a podium and condemns rioters as not "representing" the black community like they do. The social worker who has been assigned to speak for you within some labyrinthine bureaucratic system. The defense lawyer who excitedly negotiates a plea deal with the prosecutor, ostensibly on your behalf, never once imagining that you would rather do time than snitch.

Even when a representative from your affinity group speaks before the spokescouncil, or a representative from your breakout working group relays notes back to the meeting, you always feel pain. To be represented in such cases is to have not just your agency taken away but to be sliced down from all the richness and complexity of your life to a crude summary.

Hostility to this sort of "representation" is well motivated.

But to expand such feelings towards all conceptual representations of nature requires one to assume reduction or compression is *always* as impossible as it is in the messy, particularized aggregate world of humans. It's no shame for those who focus on the everyday and social instead of the scientific to not be familiar, but there are other domains in which reduction is possible and valorous. To speak of circles or electrons generally is nothing like speaking of queer people generally.

A lot of notoriously bad philosophy gets generated by poets waxing on about *"indescribable plenitude"* or *"infinite descriptions"* and the exclusions made by *"circumscription"* in ways that just sloppily retread discussions of model underdetermination, but it's a waste of time to get drawn into trying to teach poets about algorithmic complexity; so long as they feel the potential violence of political representation looming, they will squirrel out of any sincere engagement.

And certainly most self-described "science," as it shows up in many people's normal lives, does not press questions like those of fundamental physics that actually grapple with true universals, but rather intensely political and pragmatic questions at the human-scale, like AIDS research, population genetics, nutrition, and psychological pathologization.

If biologists were to declare that they had "isolated" some kind of "gay gene" or gene complexes—crudely assigning descriptive primacy to it—such a "scientific" representation would certainly bleed into political representation. The vast causal complexity of the sexualities of actual individuals would risk being stripped away before some capriciously arbitrary declaration.

It's understandable that many want to cleave away any capacity for such attacks.

EPISTEMIC DIVERSITY AS RESISTANCE TO BEING KNOWN

> "There's a philosophical objection to any refusal of universalism that will be familiar from other uses (the denunciation of relativism, most typically). It requires only one step: Isn't the denial of the universal itself a universalist claim? It's a piece of malignant dialectics because it demands that we agree. We don't, and won't ever, agree. Agreement is the worst thing that could happen. Merely assent to its necessity, and global communism, or some close analog, is the implicit conclusion."
>
> Nick Land, *Against Universalism*, 2016

When James C. Scott published *Seeing Like A State* in 1998, he introduced academia to some basic anarchist arguments and spread "legibility" as a popular term in the humanities. While Scott should be commended for digging up specific and accessible examples, all of the ideas he is now treated as the authoritative origin of predate him significantly and widely. While some postmodernists grasped some degree of it, *everyone* engaged in resistance knows the utility of making yourself illegible to the authorities as well as the desperate struggles of those in power to gain knowledge of you.

A central planner will pursue the forcible suppression of complexity so as to better manage what he controls, since all this happens at the aggregate scale of messy, inter-contingent ecology and human society, this innately creates blowback, unforeseen externalities and consequences. In the millennia of struggle between power and freedom, access or impediments to knowledge have been central; rulers try to make us legible to them, we try to make ourselves illegible to them.

Unfortunately, academic attention to this has left some circles fetishizing illegibility, and towards those ends, even rejecting reason and universalizing thought, on the assumption that irrationality somehow channels a vitalistic spirit that resists being known.

> "Tiqqun and the Comité invisible argue that, in a system that is dedicated to ever greater transparency... any resistance must deliberately seek out 'zones of opacity' by cultivating the kind of noise that won't be recuperated."
>
> Douglas Morrey, *Manifeste conspirationniste,
> Parti imaginaire, Comité invisible*, 2024

It is a grave mistake, however, to collapse legibility and illegibility into synonyms for rulership and resistance. Legibility is always *directed* and *relative*; something may be more legible to one party or actor versus another. And so too is illegibility. Locals may know their city's hive of unmarked streets better than foreign police, but this cuts in reactionary nativist directions as well:

leaving immigrants and refugees just as lost. Abusers often try to make their actions appear chaotic, unpredictable, or uncontrolled in order to keep their target off balance, under control, or even sympathetic ("he just loses control, he can't be blamed"). Similarly, a survivor who can't leave her village because the world beyond it has been structured so as to be illegible to her is less free as a result.

Just like dishonesty, illegibility is a core source of domination and social hierarchies.

It is frequently the case that people will attempt to seize or maintain power in anarchist projects by means of deliberate illegibility to newbies. An activist might perform a critical task but not make the knowledge of how to do it accessible. A role might require the laborious acquisition of tacit knowledge regarding the wider activist social landscape. The activist might suss out certain leverage points or chokepoints in flows of organizing whose very existence is highly illegible to outsiders, providing him with the ability to constrain how knowledge flows within the org, to shape who connects to who in strategic preparations or discussions before a vote, to have critical control over its representation to outsiders. This is often why newbies to activist projects consistently instinctively side with various organizationalist ideologies where formalisms are used to make things more legible. Of course this formalization often becomes a new fountainhead of illegibility that can be used to create power gradients.

The answer is not a top-down imposition of legibility, but a catalyzing bottom-up legibility. Rather than formal systems that can be captured and controlled, or who direct legibility in ways that leave thickets of illegible other things, the project of liberation is the mutual voluntary exposure of truth. We encourage norms where folks *volunteer* knowledge, and collaborate on structuring how it is presented to make it more accessible. The core value of DIY is not libertarian-style autarchy, but lowering all barriers to knowledge and action. DIY is about maximizing legibility. Indeed this is a core of insurrectionary thought: making attacks or exploits understandable enables them to be widely reproducible.

Similarly—and I can't believe I have to repeatedly argue this—everyone believing different false things about reality may in some ways increase a population's illegibility but it does not significantly impede rule over them. Sure there may be some difficulty to predict what any given person will do—since they might be just as likely to leap out of a building thinking they can fly or spend all their time masturbating to sigils hoping to levitate the Pentagon—but you can smooth over all that illegibility with the certainty that they're not going to be able to plan out effective attacks against you in the space of actual reality.

While *some* variation in focus or exploration is good, there's an early point at which epistemic diversity proceeds over a cliff and only makes resistance ineffective.

Making knowledge of how to screenprint or run a consensus meeting universal in a scene may be "homogenizing" but that is not an innately bad thing. Similarly, having a more accurate general understanding of reality itself, from trigonometry to how social power operates to how the night sky moves, enables a greater expanse of actions and options.

The reason that folks continually fall into the trap of thinking that believing random mysticism or conspiracy theories equates resistance to the powers that be is that it *feels* freeing to discard the harsh self-checks of epistemic diligence. Illusions can feel nicer than reality, but even when they don't, being able to choose between possibilities otherwise foreclosed by basic rationality is a deeply freeing experience. It's the same euphoric rush of sudden possibility as getting unchained. The problem is that it's fake. Freedom isn't a feeling, it isn't poetry or a metaphysical declaration, it's a relation in the physical world. Can you leave a cage or not? Can you eat or not? Can you load a bullet and shoot a politician or not?

But for many antirealist philosophers the structured physical world isn't as primary or essential as your raw experience.

The Fundamentalists Of Phenomenology

> "What is true is that a nature exists—but this is the nature that perception shows to me and not the nature of the sciences."
> Maurice Merleau-Ponty, *Phenomenology of Perception*, 1945

While radicalism actively *digs* for reality, fundamentalism *takes it for granted* in one's starting point or "arche." It embraces one's happenstance origin as an absolute value. Those who do not share the scientist's predilection for searching for deeper regularities and persistent firmnesses behind appearances may instead privilege every day common sense, raw sub-conceptual experience, self-reflective thought, tradition, or a holy book, because these are praised as *what we start from*.

A lot of the pushback to physicalism stems from this intense deference to one's starting point. In contrast, the radical idea that *regardless of where one starts*, there can be conclusions that any one exploring will eventually be forced to converge to is often totally unthinkable. The fundamentalist demands a foundation that is supposedly just *given*, and anything not grounded in it is suspect.

> "[Galileo] has built without a foundation."
> Rene Descartes, letter to Father Mersenne, 1638

When *The Social Construction of Reality* was published in 1966, Peter L. Berger and Thomas Luckmann were at pains to emphasize that they were not speaking of "reality" in the normal sweeping sense or as philosophers might, but merely in an informal and imprecise sense of anything fixed *independent of individual choice*, and further they were focused on the immediately noticeable staticities of *everyday life*. This came with repeated epistemic and moral emphasis that "everyday life" was somehow more primordial and more important for it.

Variations on this arche-ist instinct can be found across critics of science. From those that privilege the authenticity of the preconceptual,

> "The whole universe of science is built upon the world as directly experienced, and if we want to subject science itself to rigorous scrutiny and arrive at a precise assessment of its meaning and scope, we must begin by reawakening the basic experience of the world, of which science is the second-order expression. ... To return to things themselves is to return to that world which precedes knowledge, of which knowledge always speaks, and in relation to which every scientific schematization is an abstract and derivative sign-language."
> Maurice Merleau-Ponty, *Phenomenology of Perception*, 1945

> "Raw phenomenality is fragmented into dualisms and petrified into concepts as we self-alienate, creating frozen concepts out of flowing life. Ways of modeling and communicating our ultimately ineffable preconceptual experiences—models that are necessarily abstract and partial, revealing some aspects of actually-lived reality while they obscure others—come to be treated as somehow more real than the lived experiences from which they are necessarily always derived."
>
> Bellamy Fitzpatrick, *Corrosive Consciousness*, 2017

to those that follow Kant and Berkeley in privileging thought in some abstract mystified sense of "consciousness,"

> "If conscious experience is real, then consciousness was constructing reality all along. Having arrived at the borderline, we can look back over our shoulder and say, 'Oh, I get it now. Everything I ever thought was real is constructed from consciousness.' Consciousness isn't an add-on. It's the only thing that was real in the first place."
>
> Deepak Chopra, *Everyday Reality is a Human Construct*, 2016

These fundamentalists may disagree on precisely what is most primal, most foundational, but they agree that it comes before all that slippery and no doubt arbitrary conceptual theorizing and *induction* that might come afterwards.

Authenticity is almost always a central concern for such people, deeply bound up with notions of honesty. Influential ecologist and phenomenologist David Abram even repeated the absurd claim that the Mexicas were conquered by Cortez because their supposed greater immediacy with nature and lack of more abstract writing made them *incapable of deceit*.[110] To think deeply on one's own, out of sync with the immediate and visceral, to pick out structures that aren't already given to us, is to engage in dishonesty, including to oneself. There seems to be no awareness among such people that a reflexive response in the moment can be poorly reflective of your full self or that honesty can oblige both self-scrutiny and careful modeling of another to better convey and be received.

To such fetishists of immediacy, physics is an *abstraction*. Electromagnetism is not a *less abstract*, more firm and foundational reality, discovered after peeling away the messy arbitrary abstractions of human-scale experience. No, electromagnetism is somehow *more* abstract.

Never mind that our everyday common sense, our preconceptual expe-

---

110 David Abram, *The Spell of the Sensuous: Perception and Language in a More-than-Human World* (1996), 134. See also Leonard Shlain, *The Alphabet Versus the Goddess: The Conflict Between Word and Image* (1998) which blames abstract reasoning, writing, and radical thinking for our fall from a supposed matriarchal Eden in prehistory.

rience, our "raw senses" and "thought unto itself" are often demonstrably wildly inaccurate, with formal conceptual structures and corrective cognitive strategies developing to correct such. In the fundamentalist view, an initial visual illusion is *more real* than the "abstract" conceptual clarification that might dissolve or clarify it.

While the radical tries to find more points of contact with reality, ideally universal points of contact, and between them build a web that can more deeply press into reality, the fundamentalist tethers themselves to a single point. They believe that anything beyond that is inherently uncertain and untrustworthy, so the further they extend themselves along their single tether, the less secure they feel.

This is how you get things like Kuhn's infamous assertion that it's likely scientists in the 18th century had a *better map of reality* than the physicists of today, *because there was less theorizing back then.*

If you can't conceive of conceptual thought as a net that can inevitably close in around the only viable pattern in the experimental data, you instead see all theorizing as inherently more and more tenuous and untrustworthy speculation.

While Berkeley came to disbelieve in materialism through privileging *sensation as a mental experience*, some fundamentalists who privilege the senses dismiss *both* Berkeleyan idealism and scientific realism as deviations from *sensation as a physical entanglement with nature.*

In this view, our self-reflection and modeling takes us out of sync with the web of relations we are born embedded within. Instead of acting reflexively as one node in an ecological or social network, pushed and pulled by stimuli, and dutifully relaying such signals along, we have developed a tendency to stop and mediate on our inputs, to pull from wider nets of diverse contexts and build deeper impressions of the world around us that, in turn, allow us to see ahead and select a wider array of paths than those presented by instinct or immediacy.

This is bad, they say, because such models disconnect us from our rawest and most immediate relations. Through the "abstraction" of general thought we become Heidegger's hated enemy: the rootless cosmopolitans who reject the happenstance of identity, body, community, and land we find ourselves thrown into at birth, and choose to move beyond.

When I bring up my origins in a faith healer cult, I regularly encounter this claim that radical realists are simply the other side of the same coin as Berkeleyian idealists, both in our heads, rather than occupying our skin (often delivered in a way still imbued with trace racial connotations). Both my mother and I were born into the same cult, and both of us left. But her reasoning was explicitly sensual: she decided as a teen that she preferred sex and "the taste of an orange" over denying their existence. In the end, however, the surface had no lasting strength, and the same fundamentalist reasoning from happenstance drove her back into the cult, after all, it was where she started. The sensual offered no deep or lasting reason to move beyond her priors and community, and so they won, perpetuating abuse on another generation.

Since the fundamentalists cannot appeal to deeply to reason lest it whisk them away from their arche, the core strength of their appeals are grounded in aesthetic associations where free-flowing self-delusion, sensation, or everyday common sense is presented as obviously vitalistic, lived, passionate, raw and "human" while the radical's pursuit of depth, breadth and common regularity is framed as innately dead, cold and "inhuman."

> "Relativity and quantum physics have succeeded in doing what every branch of science would like to do; they have completely separated their sphere of knowledge from the realm of the senses... Interacting with the world on a sensual level is much too likely to evoke passion, and the reason of science is a cold, calculating reason, not the passionate reason of desire."
> Wolfi Landstreicher, *A Balanced Account of the World*, 2001

This is, of course, absurd, a complete fabrication with no contact with the lived experience of scientists. I have communed with the interplay of plants and the slow rustles of wind in the forest, the causal story they unveil and enmesh you in, and such has no special magic that is not also in the lens of a telescope or the hilarious joke of the polar integration of the gaussian function. Actual mathematics—which bears as little relation to arithmetic they teach in public schools as literature might to spelling bees—is often an explosive, imaginative, and wild arena, no less so for its structural constraints. Reason is not the cold-blooded pastiche imagined in Spock, but a passion that is searching and probing, a dance of destruction and creation.

It's particularly absurd to declare science devoid of passion or desire, when it's common to find physics students describing science among one another in very charged and passionate terms, in terms of love, empathy, and sexual desire. The closest comparison to a long effort to understand the universe better is getting to know a lover. Eureka moments of breakthroughs in understanding are often compared with orgasms, where building tension in the mind is suddenly cleared in an explosive insight that takes hold of your whole being and shakes it into a new configuration. A lot of joking in physics has historically turned on this recurring commonality between physics and sex or love. During a late night of collective problem solving while I was an undergrad, a fellow student, when she had worked out an understanding all at once, exclaimed, "*this is better than sex!*"

Now to be clear: it's my opinion that all appeals to sexual psychology are vapid and without general insight; anarchists, liberals, and fascists would all remain in a world devoid of sexual desire or repression. My point is that one can easily reverse the poetic language used to deride the radicalism of science as "cold" by emphasizing that *plumbing deeper* or taking the other *into you* has surely more resonance with the sensual and passionate than remaining limited to *surface contact*. No shame to the tugs and pulls of the surface, but why hold back from more?

What is simply *given* to you may feel comfortingly familiar, and this no doubt drives much fundamentalist clinging, but how much wild creation and discovery can one really get up to when you refuse to leave home?

But I guess there are always those afraid of how a deep love can transform you.

THE ONTOLOGICAL UPDATE PROBLEMS OF SCIENCE

> "The disenchantment of the world means the extirpation of animism...On their way toward modern science human beings have discarded meaning."
> Theodor Adorno and Max Horkheimer,
> *The Dialectic Of Enlightenment*, 1944

When Heidegger's mentor Husserl started modern phenomenology, he couldn't help but express his intention of it to be a revenge against the scientific image of the world in something very similar to Kant's notorious counter-revolution—jotting down in his notes that he had overthrown copernicanism, and that, "*The original ark, Earth, does not move.*"[111]

Such aggrieved fixations on rejecting various scientific models, as a result of their implications, are littered throughout antirealist texts.

The postmodernists in the science wars made repeated reference to various crackpot ideas they thought were clearly superior to accepted scientific models, and rejected accepted scientific models they thought were obviously false.

At the peak of the science wars, the sociologist Jane Gregory complained that scientists were ignoring the "*inconvenient truth*" of homeopathy that could make "*redundant the pharmaceutical industry.*"[112] Collins leaped to publicly defend the homeopathy of Jacques Benveniste against its total failure to replicate.[113] And Barnes, Bloor, and Henry endorsed not just homeopathy but specifically *astrology* as holding up to equivalent "*methodological principles and empirical evidence*" that would be a "*serious embarrassment to scientists if they were not so good at ignoring it.*"[114] To substantiate such confidence they cited the hilariously cherry-picked "empirical evidence" of Michael Gauquelin for a supposed influence of the planet Mars on athletic renown in humans, studiously ignoring studies that directly disproved it or broke down the statistical games being used.

We've touched on a variety of Aronowitz's declarations, since his pattern of appealing to philosophy to defend such was central in provoking Sokal, but hostility to the content scientific models wasn't always obscure or occult. Many postmodernists could be quite frank about revealing that their crusade

---

111 Peter Salmon, "After Jacques Derrida, What's Next For French Philosophy," *Aeon Essays*, May 6, 2022 https://aeon.co/essays/after-jacques-derrida-whats-next-for-french-philosophy.

112 Jane Gregory, "Reclaiming Responsibility," in *The One Culture?*, ed. Jay A. Labinger and Harry Collins (2001), 201.

113 Harry Collins, "One More Round With Relativism," in *The One Culture?*, ed. Jay A. Labinger and Harry Collins (2001), 189.

114 Barry Barnes, David Bloor and John Henry, *Scientific Knowledge: A Sociological Analysis* (1996), 141.

arose from outrage at the contemporary scientific picture conflicting with their intuitions, starting beliefs, or faiths.

Of notably recurring outrage was the reduction of the mind to the material brain and rejection of the vitalist thesis of life as a supernatural force beyond physics. These were regularly brought up in rhetorical moments of emphasis—essentially, "*and anyway we all know science is bad because it says we're mere matter!*" that reveal their intended audience. There's a narrative in sections of autonomist marxism that "the scientific worldview" arose from a capitalist push to "reduce" workers into mechanisms that could be controlled. To be sure, this is a very limited, dated, and shallow understanding of physics that collapses a lot of complexity, gets much of the causal history incorrect, and often ends up implying that the anti-reductionist organic paradigms (that embraced slavery and feudal monarchy) prior to Newtonian mechanics were somehow more liberatory. Yet, for many, this narrative is comforting because the vitalist or organicist magic of life is forever beyond scientific description and thus capitalist control, and the refusal of work becomes bound up in the refusal of science. Science thus *must* be invalidated, because anything less would allow the capitalists to win.

Such motivating concerns with the content of science was by no means exclusively leftist, it could extend to straight up christian creationism, with one of the most prominent and notorious relativist sociologists, Steve Fuller, repeatedly going to bat for conservatives:

> "For example, if (or rather when, given the fallibility of even the most indomitable of scientific theories), the theory of evolution by natural selection yields to a version of evolution that incorporates a sense of cosmic intelligence, it is certain that some historians will portray Creationists as having been prescient for highlighting the sort of anomalous evidence that today they brandish at evolutionists to little avail."
>
> Steve Fuller, *Science*, 1997

Similarly, the prominent scientific antirealist, Bas van Fraassen, was a convert to catholicism, and his derision of materialism as "false-consciousness" was directly grounded in his faith. In the broader academic background of the Science Wars, van Fraassen portrayed himself as a defender of "science," but as a mere skepticism or empiricism that elevated raw experience, while denigrating theoretical extrapolation as not even holding the tiniest sliver of potential truth regarding underlying reality. Van Fraassen's arguments found some appeal with experimentalists derisive of theoretical work in science, but to assert his bright line between "unobservable" and "observable," he was forced to endlessly struggle to secure a distinction between sensation and extrapolation, despite the two being the same dynamic in actual brains (the most "bare" neural sensation is structured as an extrapolation). He praised the sort of double-think common with modern christians who might functionally "accept" a model of reality while at the same time denying that they "believed" it. And finally he had to reject any potential change of our sensory capacities, his notion of "science" is thus *explicitly* forever tied to a reactionary, parochial, and static community that cannot be allowed to substantively grow, change, or admit minds with different faculties. Van Fraassen then justifies the god of catholicism not as a claim about physical reality that could be invalidated or tested or even compared within scientific accounts, but as a pre-scientific *felt experience*, forever beyond the critique of science.

The strategy of stripping science down to almost nothing, while leaving religion untouchable by it, was embraced almost a century earlier by the reactionary catholic philosopher of science Pierre Duhem in France. Like van Fraassen, Duhem would reach wide influence in philosophical circles for his antirealist tendencies, specifically hostility to induction and ardently holistic argument that no evidence can ever force the rejection of a hypothesis, because one is always free to tweak an assumption instead. Duhem proclaimed that science only described "*propositions relative to certain mathematical signs stripped of all objective existence*" [115] while denying the existence of atoms and sneering at special relativity.

It's easy to—pardon the act—*extrapolate* how religious commitments motivated and shaped the antirealism of Duhem and van Fraassen.

Of course such motivations don't *on their own* invalidate a resulting argument, but they can explain the stubbornness of it under pressure.

Science often demands pretty big sacrifices from our naive default models of reality. Reductionism and physicalism are hard pills for many to swallow because their most cherished values would end up having to map onto an incompatible ontology. If your present value system depends upon the existence of a god, spirits, some vital magic that sharply differentiates living things from unliving, or the notion that consciousness is somehow distinct from other cognitive or computational processes… well, it makes sense to flail against the pictures science settles on and embrace the most extreme skepticism or relativism.

All your values are directed in relation to some *things*—in the loosest sense, inclusive of structures, relations, etc. When your model of things is radically updated there is often not a clear mapping by which your old values can be applied in the new ontology.

---

[115] Pierre Duhem, *The Aim and Structure of Physical Theory,* trans. Philip P. Wiener (1954), 285.

An example most anarchists will be familiar with is the falling out between John Zerzan and the anti-civilization hyper-reactionary Ted Kaczynski.[116] Zerzan's ideological project of anarcho-primitivism was founded in large part on sweeping claims by anthropologists in the '70s that painted band societies as peaceful, egalitarian, and living with abundance. Archeologists at this time also had a very limited or incomplete record of early cities, focusing primarily on hierarchical death machines like Sumer and Ur. All this enabled Zerzan and his collaborators at *Fifth Estate* to make a relatively simple equivocation between the culture of cities ("civilization") and hierarchy, as well as between mobile hunter-gatherers and egalitarianism. But, as the decades continued, much of the anthropological and archeological evidence shifted, forced to recognize both the existence of egalitarian early cities and the commonality of violence, patriarchy, and scarcity among band societies. In the early '00s, Kaczynski—having no issue with systemic violence and oppression—cited this change of evidence and expert evaluation to force Zerzan to make a choice: was his core goal defeating civilization or was it the egalitarian aspirations of anarchism? The only way to avoid the choice was to refuse the new ontological map. And in the following years there was a pretty abrupt shift among anarcho-primitivists to deemphasize empirical data and emphasize spirituality.

Shifts of ontology often oblige shifts in values.

---

[116] Otherwise known for his work in geometric function theory.

Political And Institutional Alliances

> "Today's science is at first glance merely research into efficiency, that is, into power, and on the second merely the production of strange and efficient fictions. Not only is there no 'economic thing', there is no 'scientific thing', either."
> Jean-François Lyotard, *Libidinal Economy*, 1974

It is only natural to question what values are shaping the accounts of reality given by science. If implications are lurking that threaten to derail or hold back your moral framework, it *makes sense* to check if you're getting deliberately epistemically poisoned.

A pure mathematician by training, Jerry Ravetz came out of the leftwing science movement in the '60s, and his influential book *Scientific Knowledge And Its Social Problems* tried to codify many of the concerns the New Left had with scientific claims and "industrial society." Ravetz had a fondness for the pure ideals of science that had flourished before the war with fascism, and he painstakingly traced the increased penetration of capitalist and statist influence and incentives within science. But Ravetz's prescription was not a return to the radical realist ideal of pursuing truth for its own sake, but a concession that partially surrendered to the pragmatic instrumentalization of science encouraged by the military industrial complex, and accepted compromise with goals other than truth. Ravetz didn't disagree with the enslavement of science so much as the ends to which it was put.

This embrace of the science-as-pragmatic-engineering perspective, where one refused to speak of science on its own but rather emphasized a "technoscience" perspective, was the paradigm of the wider New Left. If science is merely a *tool*, always crafted in no small part towards some political or social ends, and the truth content of its models cannot and should not be distinguished from its political utility to various actors, then its capture by power surely invalidates or even defines the entirety of science.

"Science" is not treated as an orientation, value, cognition style and attendant accumulation of strategies, practice, tacit knowledge, and rich models, but as instead nothing more than these institutional structures, incentives, class dynamics, etc. Except the former are then constantly attacked as inexorably interlinked with the latter. The orientation is treated as inevitably prescribing the institutional rot and the institutional rot is said to give rise to the orientation as its legitimization.

This conclusion was inevitably reached by tons of leftists across the '70s and '80s, with historians and sociologists emphasizing the continuous various entanglements to the point of challenging or even erasing any notion of a distinction between science and political power. The natural progression of this institutional critique, as we've seen, was to entirely dismiss the contents of scientific models as referring to structures in reality rather than jumped up rhetorical devices to reinforce social domination.

> "Science legitimates itself by linking its discoveries with power, a connection which determines not merely influences what counts as reliable knowledge."
> Stanley Aronowitz, *Science As Power*, 1988

And to be sure, nearly all science one could point to in the last five hundred years took place within a statist society widely infested with systematic hierarchies. Very few scientists have escaped to the scant number of autonomous communal research maroons in the jungle, but even such escape or renunciation, like the famous mathematician and anarchist Alexander Grothendieck repeatedly tried, still does not achieve *neutrality* or *uninvolvement*, since that distance remains a choice with costs. And of course those scientists directly involved in active insurgent struggle are likewise always implicated in a web of power relations they must constantly strategically evaluate. There is a very trivial sense in which lines of power and the pursuit of scientific knowledge are causally entwined.

Nothing in our society can be entirely extricated from the meshed ecosystems of power and resistance. This is true of *everything*. That doesn't mean that all art, for instance, is necessarily and inherently serving the interests of our rulers. An avocado is embedded in wider networks of social power, that doesn't mean the soul of that avocados is indelibly stained with original sin, incapable of doing anything that doesn't do more harm than good in balance.

Yet the critics of the technocratic monoculture of "industrial society" were not fixated with delegitimizing art or avocados; they were incapable of seeing science as an orientation to the world, only an institutional force, one indistinguishable from existing social hierarchies.

If science valued coherence, this was surely imperialism (and probably a Jewish conspiracy to boot):

> "The idea of contradiction is an imported one from the West by the Western-educated, since 'modern science' arbitrarily imagines that it only has the true knowledge and its methods are the only methods to gain knowledge, smacking of Semitic dogmatism in religion."
> Swami Mukhyananda, *Vedanta in the Context of Modern Science*, 1997

Critics would rightly point out examples of European assumptions and frameworks being laundered as universals without any critical engagement, but refuse to see this as universalism betrayed, instead taking this as license for similar ethnonationalism and making increasingly dramatic proclamations that non-europeans, never cared for pure truth; they only wanted "ways of knowing" that facilitated the proper social order.

As always, it's amazing how deeply in-agreement such "critics" were with the analysis of the most bloodsoaked european imperialists:

> "As for the difference in philosophical perspective, it may reflect the divergence of the two lines of thought which since the Renaissance have distinguished the West from the part of the world now called underdeveloped (with Russia occupying an intermediate position). The West is deeply committed to the notion that the real world is external to the observer, that knowledge consists of recording and classifying data—the more accurately the better. Cultures which have escaped the early impact of Newtonian thinking have retained the essentially pre-Newtonian view that the real world is almost entirely internal to the observer."
>
> Henry Kissinger, *American Foreign Policy*, 1969

One must really struggle to grasp the sheer level of historical and global cultural ignorance necessary to think that realism was invented in Europe rather than something timelessly cropping up in all cultures. If anything, the European fixation after Kant's anti-copernican revolution has been to justify increasingly absurd antirealisms.

> "Truth is a vital concern of humankind across history and culture, not an idiosyncratic concern of modern white Europeans. Despite the heterogeneity of truth-pursuing practices ... A single concept of truth seems to be cross-culturally present"
>
> Alvin Goldman, *Knowledge in a Social World*, 1999

A number of the science warriors like the physicist Steve Weinberg were explicit that a defense of realism was precisely a copernican dethroning of Europe. Weinberg wrote a lengthy history of science to make very clear that it was a worldwide endeavor that far predated European advances, constantly driving home this point, *"Arab scientists in their golden age were not doing Islamic science. They were doing science."*[117]

> "What looks like a tolerant, non-judgmental, 'permission to be different' is in fact an act of condescension toward non-Western cultures. It denies them the capacity and the need for a reasoned modification of inherited cosmologies in the light of better evidence made available by the methods of modern science."
>
> Meera Nanda, *Prophets Facing Backward*, 2004

---

117 Steve Weinberg, *To Explain The World* (2015), 123.

WHAT WAS CLAIMED TO MOTIVATE THE SCIENCE WARRIORS

Social Text repeatedly tried to unite around the narrative that what the science warriors were really objecting to wasn't antirealism but any and all "social analysis" of science, its history, or really anything that was leftist or critical of the institutional and political power dynamics. This narrative, of course, could only get off the ground by entirely ignoring and hiding that the science warriors continuously had praised sociological and historical accounts of science and even the more conservative ones been quite forthright that, "*science is, in some sense, a cultural construct,*" to use Gross and Levitt's phrasing.

Social Text howled about civility and respect they were owed as academics, but the counter-narrative they pushed the hardest was that objections to postmodernism were nothing more than the desperately misplaced financial self-interest of scientists, picking fights with the humanities to somehow steal back lost funding.

> "Government is decreasing its funding of science and requiring greater accountability, and scientists, often working in the interest of private profit, are facing increased difficulties of self-regulation. In this context, "outsiders" who study science are convenient scapegoats, and waging war is an easy way for scientists to avoid critical self-inquiry, to deflect responsibility and blame."
>
> Dorothy Nelkin, *Social Text*, 1996

Or, as Babette E. Babich alleged the scientists thought in *Physics vs Social Text*, "*Nip the 'bad' writing of deconstructionists and postmodernists in the bud and funding levels for supercolliders will be up in no time.*"[118]

The postmodernists repeatedly referred to the science warriors as "*the barbarians at the gate*" with all kinds of allusions to moral and—equivalently—cultural stuntedness. They took it for granted that *only they* had conscious moral or political motivations, whereas the scientists were obviously naive, ignorant, little bureaucrats whose surprising fervor could only be explained by conservatism and selfish institutional ties. *They* wanted to examine social power dynamics, to help people and resist domination, whereas the scientists criticizing them were self-evidently spasming in unthinking defense of a technocracy without any sort of driving concern for liberation. What else could you possibly expect from scientists who didn't even study people or medicine?

If natural scientists cared about other people, they wouldn't be scientists.

In academia's two cultures war, the humanities' desperation for funding and respect often turns on a narrative where "the humanities" are some kind

---

[118] Babette E. Babich, "Physics vs. Social Text: Anatomy of a Hoax," *Telos* 1996, no. 107 (1996): 45–61.

of natural domain in knowledge defined by caring about people and acquiring eudaimonia, the good life. In this way "the humanities" are made functionally synonymous with *ethics*. By the magic of interdisciplinary coalitions, studying Shakespeare is the same thing as studying anthropology which is the same thing as studying art history, and, in contrast, studying mathematics is the same thing as studying climate science which is the same thing as being a bitcoin bro.

While in a lot of humanities it's usually considered gauche to speak of ethics *directly*, as that might involve engaging in rigorous analytical philosophy, it's totally appropriate to gesture at ethics *nebulously*, implying that just studying anything within the humanities bootstraps a conscience and awareness into you, whereas the cold inhuman calculations of the STEM janitors could never. (If this were true there'd be a whole lot less rapists in women's and gender studies departments, and any anarchist or feminist with any experience can tell you that's not even remotely the case.) What this narrative turns upon really hard is a kind of loose impression of a Humean is-ought-gap. The humanities are therefore somehow all about what *ought* to be whereas those in the sciences merely teach what *is*.

This narrative only gets any traction because there's already a widespread conflation of sociality with ethics.

One of the most popular confusions you run into as an activist organizer is that a huge amount of Leftists think *"doing things alone creates selfishness; doing things together creates compassion."* This is obviously not even remotely true, as collective decisionmaking in no way changes the motivations of those involved, it merely shifts the dominant flavor and styles of power dynamics, with many continuing to ruthlessly pursue self-interest and control. But this belief persists as an instinct shared by every new generation of activists—until they burn out—and it seems to be a general phenomenon in our society. A huge chunk of the population fervently believes that egoism vs. altruism is the same thing as introversion vs. extroversion which in turn is the same thing as primary interest in studying matter vs. studying society which is the same thing as analytic thinking vs. poetic thinking. Everyone involved recognizes this is absurd when pointed out, but then they constantly default on it or leverage it aggressively *anyway*.

Broader Scientist Moral Narratives

While the narrative of the sciences as amoral and the humanities as a bastion of morality is ubiquitous in print and a staple of academic posts on social media, there is a *second narrative* that I rarely see referenced explicitly in print and that many of the humanities crusaders seem totally unaware of, but one that I've found widespread throughout physics and math departments. It's almost the exact inverse: *studying social matters is something that pretty much only those seeking power and domination would do.*

> "When a scientist says something, his colleagues must ask themselves only whether it is true. When a politician says something, his colleagues must first of all ask, 'Why does he say it?'; later on they may or may not get around to asking whether it happens to be true."
>
> Leo Szilard, The Voice of the Dolphins and Other Stories, 1961

We might see this as a difference in personality that forms early. Two groups of kids form up: one plays mostly *in parallel* with rocks in the creek and the other plays mostly *with each other*, by for example playing tag or gossiping. To a number of the kids whose play is externally directed at the material world, *the social* has been revealed as a grotesque and horrifying place of manipulation and power plays. There's no point in studying these dynamics because they're a ladder of arbitrarily escalating complexity in a zero-sum game. The purely social world is a closed system, with no positive inputs, so all it can consist of is pointless status competitions. Whereas those who are externally focused on the material world can *avoid competition and domination*, because each kid digging a hole or stacking rocks involves the domination of no other agent, just extending agency over inert matter.

In this picture the valences are stridently reversed: to do things collectively/socially is seen as inherently selfish. To study people is inherently suspect if not only explainable by being a manipulator in pursuit of status. I've been critiqued endlessly by well-meaning hard science nerds throughout my life for being at all engaged in social and political issues (to say nothing of organizing or writing), and the moral outrage usually takes the form of saying that *only abusers, tyrants, cultists, and cops would have such an interest.* I've lost count of the number of anarchists I've watched leave the movement after years of activism, despairing that attempts to solve social problems through social means are incoherent because the social is all selfish power dynamics all the way down, and if you really want to help people your best bet is to focus on soil regeneration or encryption instead. *Good people avoid manipulating one another by avoiding thinking in too much depth about social relations and instead focus on thinking about / manipulating inert things.*

While this is all just the inversion of the cartoonish narrative pushed in the humanities and is trivially incorrect, I bring it up not merely as a counter-narrative with at least some insights to consider, but to show that while the modern conflict between the humanities and sciences is often painted as one between the morally righteous and the *apathetic* (if not complicit), it is at least as often between two morally righteous narratives.

At the same time this general concern with manipulation can be quite valid, especially when it relates to concrete physical facts and people's maps of them. One of the primary concerns folks often have with "social construction" is not that for instance the rules of baseball are trivially constructed and could be otherwise, but rather that it opens up our models of physical structures (like whether Jake pulled the trigger on the gun) to infinite mutability via malicious strategies of negotiation. Everyone over the age of three has rich experience with how social consensus can be shaped by bravery contests (e.g. games of chicken), misdirection in language (e.g. obscuring certain details out of existence by making them harder to express), as well as dynamics of charisma and politicking around democratic majorities, legitimization hierarchies, and boycott coalitions.

But it's especially important to note that changing social consensus away from reality is often *mutually beneficial* to all parties at a given moment. Until the invention of writing and permanent physical signification started interfering with it, one of the free parameters in peace talks between conflicting parties was always the consensus account of *what had happened*. In this way wars could be defused by both parties agreeing on a historical revision. Everyone is invested in the new "reality" for the sake of peace and so the new "reality" gets bootstrapped as consensus (minus maybe a sacrificial minority).

In this light, stubbornly holding onto belief in a fixed objective reality outside the social makes one *a threat to the social fabric* and almost every other participant in society. The physicalist is, in essence, deeply and viciously *anti-social*. They elevate something *outside of society* above society and partially suborn themselves to it. This altruism of *epistemic* rationalism (i.e. pursuing the truth for its own value) makes them, from the perspective of society which operates on more pragmatic *instrumental* rationalism (i.e. pursuing what works), violently irrational. We are not amenable to pressures upon our self-interest (or pressures to adjust our relevant interests) and so are not good citizens or participants in the game of power but rather get viewed as almost roadblocks, indigestible irritants.

The Leftist Science Warriors

> Sokal certainly relished his role as an irritant.

> "Why did I do it? I confess that I'm an unabashed Old Leftist who never quite understood how deconstruction was supposed to help the working class. And I'm a stodgy old scientist who believes, naively, that there exists an external world, that there exist objective truths about that world, and that my job is to discover some of them."
> Alan Sokal, *Transgressing the Boundaries: An Afterword*, 1996

However he was characterized—as an undersexed neurotic, etc.—he would take great delight in accepting the description. But there's clearly some truth to attempts to pigeonhole him as just one more of the stodgy marxists who filled the '80s with tracts bemoaning the rise of postmodernism. Many of these unreconstructed troglodytes complained bitterly about the rise of liberation struggles around gender, race, and sexuality that did not position themselves as mere derivatives of capitalism's underlying economic exploitation. Most of these critiques are clearly more horrified by anarchistic influences in corners of postmodernism—*decentralization? complexity? a microphysics of power? in my leftism??*—than they were by any antirealist moves.

It's not clear to what degree Sokal was influenced by or sympathetic to these criticisms.

More telling, I think, is that while his personal horror at antirealism and relativism comes off quite intense and sincere in everything he wrote, he also tended to quickly skew off into critiquing standards of thought and discourse—of *communicative rationality*—instead.

> "What concerns me is the proliferation, not just of nonsense and sloppy thinking per se, but of a particular kind of nonsense and sloppy thinking: one that denies the existence of objective realities, or (when challenged) admits their existence but downplays their practical relevance. At its best, a journal like *Social Text* raises important questions that no scientist should ignore—questions, for example, about how corporate and government funding influence scientific work. Unfortunately, epistemic relativism does little to further the discussion of these matters."
> Alan Sokal, *A Physicist Experiments With Cultural Studies*, 1996

When he published *Fashionable Nonsense* with Jean Bricmont, their target was not antirealism but sloppy language, obscurantism, and misuse of scientific concepts. Only much later did he publish *After The Hoax* attacking philosophical antirealism directly. It's clear from this that Sokal's theory of social change puts pride of place on clear thinking and plain language; in short: *we* must sort out a plan to bring down capitalism, and this is imperiled by tangled metaphors and bad arguments.

Far more strident were feminists and younger scientists like Meera Nanda, for whom science remains *the* site of liberatory struggle—*"Modern science is the standpoint of the oppressed"* [119]—because domination continues to be critically dependent upon eroding or impeding accurate knowledge, informed agency—*"Superstition is as much (if not a greater) a source of tyranny as any despotic power."* [120] For Nanda, the struggle against patriarchy and caste domination was inseparable from completing the millenia old atheist struggles in India and breaking the back of hinduism. Science is not a matter of proper discursive etiquette for her, but a radical political project of physicalism to *"desacralize"* and *"demystify."* Not a staging ground from which to collectively plan attacks on capitalism, but a blade to be thrust into the chest of reaction.

This is a model of domination that posits its primal source not in differences of physical capacity or managerial aptitude, but in various strategies to stop people from knowing. Truth is thus partially *constitutive* of liberation.

Nanda's origin as a science warrior was the liberation she found in science's far more accurate models of reality that stripped away the confusion and cruft that traditional patriarchal authority leveraged. And consequently, the shocked betrayal she felt as a young scientist to discover that titans of the Indian academic left were sneeringly aligned against this liberation. Worse, they were often in bed with the very fascists she had escaped from.

While much of the rest of the Indian Left had written off traditionalists and nationalists as a dying enemy, already consigned to the dustbin of history— and thus an occasionally useful ally one didn't need to worry about—Nanda spent the '90s raising the alarm about hindufascism.

Just as philosophical antirealism was reinscribing authoritarian frameworks, the relativism and orientalism of postmodernism was building new legitimacy and prestige for old despots. Particularism and anti-reductionism were not new and novel concepts in India, she begged white leftists to understand, but at the foundations of ancient systems of oppression that are in no sense extinct.

> "What appears as marginal from the point of view of the modern West, is not marginal at all in non-Western societies… The right wing can use the same logic and invoke the same local knowledges much more 'authentically' and 'organically' for it can mobilize all the traditional religious piety and cultural symbolism that go with local knowledges."
>
> Meera Nanda, *Prophets Facing Backward*, 2004

---

119 Meera Nandra, *Prophets Facing Backwards* (2004), 203.
120 ibid, 80.

A Personal Tale

> "The philosopher is born of a single question, the question which arises at the intersection of thought and life at the philosopher's youth; the question which one must at all costs find a way to answer."
> Alain Badiou, Preface to *After Finitude*, 2009

As previously mentioned, my mother subscribes to the very common belief in the US and India that thoughts constitute reality, that physical conditions are the direct product of faith. If you want money, believe that money will magically be deposited in your bank account, and reality will warp and contort such into existence. If your leg is broken, simply believe it back into wholeness. If your leg still persists in being broken the problem is either your lack of faith *or* it is a product of others maliciously choosing to maintain their belief in the broken bone.

This latter sense of the *social* construction is critical. My mother fought hard to avoid vaccinating me and my sisters as children, only relenting in the face of severely coercive legal pressures. This resistance to *"the materialist medicine path"* was motivated by a compulsion to resist the totalitarian hegemony of belief in science and materialism. To comply, to participate in the rituals of the adversarial perspective on reality, would be not only to risk eroding one's own certainty in the non-existence of matter but also to *affirm* the majority's ongoing belief. And the more overwhelmingly others in society believed in this material world, the more actual force it would develop, the more it would be *actualized*. Going to the doctor was complicity in the popular belief that sickness was even possible. Getting a vaccine against a disease *would make the disease more real*, increasing the number of people deluded into being injured by it.

While my mother wouldn't have put it in these precise terms, she very much saw us as engaged in a grand psychic war to undermine the stranglehold of a hegemonic empire of science. The stakes were everything. If science won, then disease and gravity would be real. But the more people came to believe otherwise, the less such constraints would exist. People would fly like Superman, turn water to wine, bodies would dissolve into pure mind, and eventually we could unmake the illusion of material reality altogether... so long as we could wrest ourselves free from the tyrannical grip of material science.

It should not be surprising that believing your thoughts entirely construct reality is a highway to abuse. Feeling mad? Brutally take it out on your children and then—when you feel a touch of guilt—you can *literally rewrite history* by believing it never happened.

The only hitch is if your children (or the material evidence) persist in claiming the thing you believe never happened did in fact happen. Such infringement upon your autonomy to believe what you like and live in the resulting personal reality is obviously *aggression* and *oppression*. Further it's clear on a little reflection that such violations of your epistemic sovereignty can only be motivated by malice! If you briefly believed that you had kicked your daughter's bedroom door off its hinges or thrown your daughter out of a car on the highway—before correcting such a belief—what would motivate that daughter to try to reinstate the discarded non-reality? As our mother screamed in outrage several times a week our entire childhood, *"What would it benefit you to believe that happened?!!"*

The answer, of course, was that she had contrived so as to make it benefit us very little to believe in the existence of a non-fungible external *objective* reality. Even if our priority was to preserve some sense of *autonomy*, the more useful path was to wall off our own epistemic universes, believing our own realities, constructing our *own* pragmatic and self-assuring lies in contrast to our mother's. To reject the lies that were *useful* to our daily survival and instead focus on what was *true*, no matter the destructive cost that imposed on us, was, in a significant sense, a defiant self-abnegation.

The watershed of liberation in my childhood was obtaining a secret *tape recorder*. A blunt material object that, as it turned out, recorded *only one* version of events when surreptitiously smuggled into the presence of my mother's rage fits.

What was so potent about the tape recorder was not that it disproved my mother's claims, but that it could *surprise* the listener. After all, the tape recorder might just be an artifact of my own *"mental malpractice"* and *"manifest character assassinations"*—just a material congealing of my own beliefs about what had happened. But whenever I got gaslit and sufficiently beaten down into sincerely believing that my mother had never done a thing or had never reversed opinion two sentences in… the tape recorder *would disagree with my own memories and beliefs*.

If all matter was simply a product of our personal perspectives and a rough or localized social consensus, if physical reality was not a *thing* but pure social construction, it should not be able to *push back* like that. If no one in existence believed that my mother had done something, there was thus no mental force that could create material evidence that she had. Much less—when there was a multitude of evidence—for it all to line up.

There is, in fact, only one reality. Only one truth.

Access to it may be limited in various ways, but in the grand geopolitical arena of the family living room there are no fractured pluralistic regions of "different truths" split up between the empire and the subaltern. Nor is the empire's crime one of impeding or intervening on the subaltern's *own* reality-creation. The real epistemic violence of the empire is in cutting off, misdirecting, or impeding the subaltern's access to *objective reality*.

One of the core roots of domination is in degrading the accuracy of another's map of the objective conditions of reality and thus one's choices and even capacity to formulate the process of evaluation in choice itself. Liberation is not self-creation, but accuracy. Because delusion, whether self-inflicted or externally-inflicted, still constrains your capacity for action.

Similarly, victory is not a matter of the empire believing one thing and the subaltern another, each pickled apart in epistemic apartheid—their museums proudly proclaiming one history, while yours angrily denouncing a different one. There can be no decolonization that leaves the empire bubbled off in its delusions. Until they are forced under the singular universality of *actual reality* to recognize and understand their atrocities, *"liberation"* is just the empire's comfy retirement plan.

The singular universality of reality matters. *Is* the nature of matter.

As a result of how the two cultures have since played out, it's often forgotten that one of Horkheimer's core critiques of positivism was that it was, in his opinion, *antirealist* about science. He worried that mere empiricism with its humble unwillingness to assert the existence of root dynamics like electrons, settling only for an account of surface or apparent relations, was leaving science weakened when it came to rejecting superstition as well as correctly analyzing social issues. His classic example is that a bird that has had its vocal cords severed may not cry out, and so at the surface level, appear indistinguishable from a bird not in pain. (Neurath quibbled that such surgery could be ultimately detectable, a pretty dorky refusal to engage with a thought experiment, but at least he got the zinger in that the intellectual scam of *dialectics* was certainly not going to do a better job detecting such surgery.)

Again and again, throughout the texts of the core science warriors you find these same political concerns and critiques. *"Liberation"* is a consistent watchword, even among the less leftist and more liberal warriors, and the pressing questions returned to are not just how can we escape from under the boot of religion and tradition, but how can we even have freedom at all without accurate maps of reality to inform our agency.

These are boring and plain points, I know, cringeworthy and unfecund for interpretive exploration. This is the sin of truth: it's often incredibly boring and obvious and also *incredibly important*.

In later sharply breaking with the mainstream of STS, historians within the field would evaluate that Bloor's project *"denies the public character of factual knowledge about a commonly accessible world"* and that, decades on, the results were clearly in:

> "The winner of this particular 'game' is almost always status quo power: the conservative billionaires, fossil fuel companies, lead and benzene and tobacco manufacturers and others who have bankrolled think tanks and 'litigation science' at the cost of biodiversity, human health and even human lives... It is a sad irony that STS, which often sees itself as championing the subaltern, has now in many cases become the intellectual defender of those who would crush the aspirations of ordinary people."
> Erik Baker and Naomi Oreskes, *It's No Game: Post-Truth and the Obligations of Science Studies*, 2017

And this is almost verbatim the sort of critiques and concerns regularly howled by leftist physicists in the science wars. There is a single reality and efforts to obscure or reject that are always inevitably a boon for oppression. Truth is not some arbitrary byproduct constructed by power, but rather power is in no small part built on getting people to hold *less accurate* models of objective reality.

As countless figures like Charles Mills emphasized during all this, one of the core ways that *actual* authoritarian regimes function is by eroding the populace's epistemic traction on reality. This can take place via injecting disruptions in the networks by which one interacts with the wider world, so that you are rationally forced to degrade your credences and waste cognitive resources considering a vast chain of alternative possibilities. Newscasters pushing naked propaganda, coworkers squirming to say whatever they feel incentivised to, compromised devices, filtered archives, poisoned data, firewalls. But authoritarian control can also take place by encouraging an ethos or norm of undue generalized skepticism. Oppressive regimes are utterly pervasive with wild conspiracy theories and a general malaise or lack of care around epistemics. When you provide strong critique or counter evidence to a street vendor echoing some absurd conspiracy theory he shrugs dismissively. Everyone knows that all theories are equally absurd, so why bother ever adjusting your implicit distribution of credence in response to evidence, coherence, etc. This is *more* than the rational consideration of multiple possibilities; it is an unwarranted skepticism that innately elevates the false in dismissing pursuit of the true.

When Jane Gregory complained that scientists were (by following basic Bayesian reasoning) ignoring the *"inconvenient truths"* of homeopathy she was defending as plausible the same "water memory" scam that was literally cited by my church teachers as proof that material reality is the product of our thoughts.

I cannot express the depths of betrayal I felt as a teen, having escaped from a prison of antirealism only to find many of the exact same arguments that underpinned my oppression circulating in the Left and even in some corners of anarchism. The people who were supposed to be my comrades in the war against all power turned out to be fervent defenders of the boot I'd grown up stomped under.

This is an overwhelming and visceral fury that Nanda and I share, and that will burn and fester in our guts until our deaths.

Both Nanda and I grew up in spaces where pluralism about and even outright denial of a physical reality independent of our minds was hegemonic. Neither of our contexts are considered valid concerns. There are institutional incentives in academia to avoid *uninteresting* problems, but there are also class reasons that our lives and concerns are ignored. The more consistent antirealism of faith healers is dismissed out of sight in the halls of academia not *in spite* of its widespread popularity and influence in places like the US and India, but *because* of it.

When pressed by a few attendees of the seminars that became *Pragmatism As Antiauthoritarianism*, Rorty twisted around in some awkward distress, asserting that the answer to things like faith healing was just to *live in a community where the consensus is that that's wrong*. Why? *Well because we've accepted that for whatever particularized reasons*. But how can one justify *or even* conceptualize a *categorical* rejection of faith healing that isn't likewise an assertion of realism? Rorty's particularist pragmatism has no capacity to answer, and he has no desire to; his fixations are elsewhere.

Just as Foucault treated the children sacrificed to the pederasts of ancient Greece or women sacrificed to the mullahs of Iran, those under the boot of faith healing traditions that embrace antirealism are mere *friction* or *collateral damage*—if we are even to be considered—in the struggle against the near enemy. We are beyond the periphery of rhetorical concern; we are in the sacrifice zone.

> "The mother of a young boy who died from untreated diabetes, and the Christian Science practitioners who administered prayer rather than insulin as the 11-year-old slipped into a coma, failed today to persuade the Supreme Court to hear their appeal of a $1.5 million damage judgment won in a lawsuit by the boy's father."
>
> *The New York Times,* Christian Scientists Rebuffed in Ruling By Supreme Court, January 23rd, 1996

The Stakes of Antirealism

"Sokal is attacking a view that no-one holds. Except perhaps the late Bishop Berkeley."

McKenzie Wark, Physicist Opens Fire in the Science Wars,
The Australian, May 25th, 1996

6

THE ANTIREALISM WITHIN THE HOUSE

"Some physicists would prefer to come back to the idea of an objective real world whose smallest parts exist objectively in the same sense as stones or trees exist independently of whether we observe them. This however is impossible."
Werner Heisenberg, *Physics and Philosophy*, 1958

"We now know that the moon is demonstrably not there when nobody looks."
N. David Mermin, *Boojums All the Way Through*, 1990

"I am certainly not an atheist or a conceited agnostic."
Paul Feyerabend, *The Tyranny Of Science*, 1992

In the early '90s, just before the Science Wars grew to widespread attention, Feyerabend learned he had contracted cancer. Instead of quickly pursuing medical care, he went to a faith healer.[121]

In my 20s, while squatting in the UC Berkeley physics buildings or hanging with old leftist and anarchist physicists, I would intermittently hear stories about the man that had tried to destroy science. *Paul*, they would say with a sigh, *was one of us. He was just a troll, a provocateur. He didn't mean most of what he wrote. He didn't know they'd take it so far. Yes, he went a little overboard. We all agree he was right about there being no precise formal "scientific method" as such. He just expressed his point with a bit too much rhetorical pomp. He didn't actually think that physical reality wasn't real or that anything truly went when it came to finding answers. Those were just dramatic overstatements of his case. He was one of us, after all.*

Against Method originated as part of a collaborative project with Feyerabend's friend, Imre Lakatos, wherein each would argue opposing sides strenuously, a kind of staged performative debate. But Lakatos died and Feyerabend decided to publish his half stripped of its original context and counterweight, an extremity dangling in the wind.

As the polemic became a runaway hit for Verso and his lectures at Berkeley were packed with young leftists, Feyerabend played to this audience. He learned what lines drew applause or interest, and delivered them with relish. Meanwhile, with his physicist friends, he took a different tone, playful and dismissive of anyone taking his words seriously. He was just a cartoonish rockstar, and rockstars needed to release albums. They needed larger than life provocations.

When he doubled down on his declarations that science was no better than faith healing or astrology by attending practitioners of both, few took it as anything other than a bit. What damage could such showy endorsements do? Who could take him that seriously when he brought creationist, wiccan, and faith healer friends into his classes to teach?

Meanwhile, up a long winding driveway in a Portland suburb, my Sunday School teacher triumphantly pointed to another article in a popular magazine that validated the church's belief that matter did not exist independent of the

---

121 John S. Wilkins, "How Not to Feyerabend" *Evolving Thoughts,* October 5, 2007 https://evolvingthoughts.net/how_not_to_feyerabend/

mind and that reality was nothing more than faith. *Scientists and philosophers agreed with us!*

In the big room where all the grown-ups went, the church organ groaned to a stop, and it became time for the week's testimonials where each member would stand up and pontificate about how they had changed reality with the power of belief this week, from solving office dramas to knitting bones back together in a single instant.

My Sunday School teachers weren't the only ones who ended up taking Feyerabend seriously.

He died on February 11, 1994.

A decade later, in 2004, the film *What the Bleep Do We Know!?* premiered at the Baghdad Theater in Portland, Oregon. It's an old indie theater on Hawthorne, across the street from Powell's Books' second location. At the time, gentrification hadn't finished ravaging the neighborhood, a ton of cramped secondhand shops lined the street while hippies and punks sold trinkets on blankets. I had been insistently invited to the premiere weekend by my mother; she had heard good things and was certain this would help convince me of her faith. Dennis Kucinich canvassed for votes in the lobby, he was running for President in that year as the candidate of the American far-Left, and there was no greater concentration of his voter base to be found in the country.

The film—infamously loathed by every physicist—argues in a friendly cartoon-riddled kind of way for the widely held perspective that matter has no separate stable existence, but is the product of our thoughts; that belief creates reality. Quantum mechanics, *as everyone knows,* proved this. The film interviews two former physicists Fred Alan Wolf, by then a professional shuckster of New Age drivel, and David Albert, who said his interview was deceptively edited (it pretty obviously was). I decided about 15 minutes into the film that every single person who had deliberately bought a ticket deserved to suffer criticism and fact-checks.

To many, these criticisms were a complete surprise. Everything that the film said was practically common sense. My mom's church friends, including one a retired leader in the CIA (a number of her cult were involved in Iran-Contra), were particularly hurt and aggrieved that I disliked the film. *Just finish your physics degree,* he urged, *then you'll see.*

> "There is no deep reality."
>
> Neils Bohr, apocryphally

In the 1920s a small band of physicists came across an impediment: It seemed impossible for experimental apparatuses to simultaneously measure the momentum and the position of a particle at the same time. Comparable impediments had occurred before in physics, but at scales where greater resolution was eventually possible. It's easy to understand that if the only way you can interact with a billiard ball or get information about it is by hitting it with another billiard ball and causing it to alter trajectory, there will be certain tradeoffs in what you can learn. But this time out of thin air this band of physicists made an audacious philosophical proclamation: there was no deeper structure to be found. The theory was not in some sense immature but rather entirely complete.

The physicists who together congealed on this interpretation were a motley crew, some influenced by the antirealist wing of the logical empiricists, others by a need to legitimize physics to the mystical and romantic currents popular in German society at that time. Heisenberg would eventually help lead the Nazi nuclear bomb program, but until the end of the war he felt constantly under siege by widespread German antipathy towards physics, quantification, and scientific realism. The claim that *we cannot know*, started to slip towards the claim that *there is nothing to be known*, and then towards the claim that *there is nothing besides the knower*, or at least that some conscious knower is critically involved. They weren't quite kissing the occultism that was gripping Germany, but they could at least get close.

Many of the physicists in this small circle were inconsistent in their speculative writings, but worse, many adopted and reiterated language haphazardly, leading to talk of "observer" when nothing whatsoever suggests that certain dynamical novelties require a conscious mind to be involved in any way. Nothing, *absolutely nothing*, in physics even begins to suggest that the interference pattern of the infamous double-slit experiment wouldn't exist in the absence of a mind to observe it. And yet Heisenberg was excited to chatter about it in a way that gave license to mystical and idealist interpretations.

Einstein was infamously a hardline realist and represented much of the wider physics community who rejected the antirealism of the positivists (to say nothing of the mysticism) around the Copenhagen interpretation. This is often framed in popular writing as a failure of an old man, destined to be replaced by an incommensurate paradigm he was too limited to accept. But actually nothing of the sort happened.

Today almost every major contemporary interpretation of quantum mechanics is realist. Pilot-Wave theory (which predicted particle/wave results in 1923 before their observation in 1927) is often brought up as realist because it defends some common intuitions, but so too are the popular interpretations of Objective Collapse theories and the Everettian multiverse. There are fair arguments for each, given present limits to evidence. The newcomer Quantum Bayesianism has had some troubles and is not comparably popular, but many of its varying formulations are realist or at least passionately profess to be; in their view, the wavefunction may capture only our knowledge, but it still reflects a deeper physical reality we have no say in. And the modern advocates of what remains of the Copenhagen zeitgeist tend to be pragmatic instrumentalists who in no remote sense suggest that our minds or "consciousness" have any role in the constitution of quantum level reality. The problematically evocative term "observer" was always easily replaced by "physical experimental apparatus."

In short, nothing discovered in physics looks even remotely like the claims that *belief* makes electrons spring into existence or your *emotions* make water molecules reshape themselves that *What The Bleep Do We Know?* enthusiastically repeated.

There remain significant divisions within physics between a fraction of instrumentalists who advocate extreme skepticism of theories, and theorists who—quite naturally—believe the whole fucking point of our theories is to find purchase, however partial, upon the underlying structures of reality. When students are taught about the complex amplitudes in quantum mechanics (i.e. the "strange" dynamics of probability and particle self-interference), the canned explanation is that objective physical reality is under no obligation to match the intuitions and daily experiences of humans; indeed this is parsed as justification for realism, it would be *weird* if the deep structures of the universe *weren't* strange to bipedal primates.

Sure, the lightly antirealist legacy of logical positivism retains some purchase among experimentalists. But almost no one besides the most marginal of cranks and grifters believe "observers," in the sense of conscious minds rather than physical experimental interactions, play any role. Indeed there are very sharp and devastating arguments against such in philosophy of physics relating to different types of absurd arbitrariness necessary for such an argument.

Yet to read postmodernists like Aronowitz in the '80s, they seem highly confident that the arc of history in physics is towards precisely this centrality of the observer and the disproof of physicalist realism. Dwelling primarily on texts from the first half of the century they remained astonishingly unaware that these sentiments had little support among physicists rather than just occultist zinesters. (And, to be fair, Feyerabend, who was perfectly aware of this decline and the reasons for it, yet still peppered his public lectures with reference to "the observer" under the guise of merely *teaching the controversy*.)

For their part, the military industrial complex was more than happy to push its thumb on the scale to buoy shallow empiricism among the instrumentalists, explicitly hostile to theoretical physics and probing realist or radical questions. This in turn facilitated sliding from Antirealism5 to Antirealism3, from the faux humility of *"we can't know reality"* to *"we create reality with our minds."* Mermin, whose cringe quote on the nonexistence of the moon *when we don't look at it* graces the start of this chapter, not so coincidentally played a major role in codifying the slogan *"shut up and calculate"*—the ideological line of the Manhattan Project and the modern military industrial complex towards physics. It should be unsurprising that the state was deeply invested in trying to turn physics from a science into an engineering discipline. From radicalism to pragmatism.

But it's not like the public invented their mystified notions on their own. Part of the reason Sokal and company were so aggrieved by the opportunistic circulation of misinterpretations of physics within the postmodernist movement was the horror anyone would feel at those among one's own ranks being used to justify atrocity.

One can and should condemn the careless approach that a select number of physicists have had towards questions of realism and antirealism. Yes, many of them were young, Heisenberg, for instance, published on the misleadingly named Uncertainty Principle at age 26, and it is a bit unfair that slapdash notes and letters fired off amidst heady discoveries can be used by opportunists in later generations as justification for faith healing.

But youth is no excuse; there are important stakes at play and they always deserved serious consideration. We cannot afford to avoid criticism of those among us who fall short and do so to grave impact on the wider world. We cannot afford to be neutral or removed from struggle.

While I spent my teenage years as an anarchist organizing protests and fighting the cops, Feyerabend spent his teenage years in the Hitler Youth, by his own account with a generally flippant approach to everything, sometimes dragging other kids to nazi meetings and feeling no concern for social issues like the removal of the Jews.[122]

> "A Dadaist is convinced that a worthwhile life will arise only when we start talking about taking things lightly and when we remove from our speech the profound but already putrid meanings it has accumulated over the centuries ('search for truth';'defence of justice'; 'passionate concern'; etc., etc.) ... I hope that having read the pamphlet the reader will remember me as a flippant Dadaist and not as a serious anarchist."
>
> Paul Feyerabend, *Against Method*, 1975

---

122 Paul Feyerabend, "Occupation and War" in *Killing Time: The Autobiography of Paul Feyerabend* (1994), 36-53.

7

JUST FOUND OUT ABOUT OBJECT PERMANENCE

"What we've encountered, I think, are the limits of pluralism or the view that everything is just a story or narrative... Someone might ardently believe in faith healing, but their child still dies from meningitis."
Levi Bryant, *More Remarks on Pluralism*, 2014

"How can [science] depend on culture in so many ways and yet produce such solid results! Most answers to this question are either incomplete or incoherent. Physicists take the fact for granted. Movements that view quantum mechanics as a turning-point in thought—and that includes fly-by-night mystics, prophets of New Age, and relativists of all sorts—get aroused by the cultural component and forget predictions and technology."
Paul Feyerabend, *Atoms And Consciousness*, 1992

"Inquiry in the sciences is like empirical inquiry of the most ordinary, everyday kind—only conducted with greater care, detail, precision, and persistence, and often by many people within and across generations; and that the evidence with respect to scientific claims and theories is like the evidence with respect to the most ordinary, everyday claims about the world—only denser, more complex, and almost always a pooled resource."
Susan Haack, *Defending Science Within Reason*, 2007

Perhaps because I was raised from birth to believe that the physical material universe does not exist at all, with literally none of its perceived regularities being anything but—in the words of the church—absolute "error," I cannot help but judge weaker antirealisms in terms of their relations to stronger ones. When the full resources of a position are devoted to skepticism in one direction while it mobilizes next to nothing to fend off another position, what functionally distinguishes them?

If you strenuously reject that we can have any inductive knowledge of the world while appending a statement that obviously some external physical material world no doubt exists, why the hell should you or anyone else actually accept that extraneous claim?

The notion of object permanence that an infant develops is a hypothesis, an extrapolation of deeper roots under the blinding buzzing confusion of sensation. From the start of our lives we search out ways—patterns—by which to compress the apparent chaos as much as possible, and then compressions of these compressions. We test all of this from all directions, and we throw out the compressions that fail to have the same scope. We slice the world up into potential root realities, objects, and grow so confident in our slicings that when some objects are out of sight—when we have no "direct" perception of them—we nevertheless come to embrace the proposition that they exist. Baby's first radicalism.

One could not ask for a more prototypical scientific hypothesis than this object permanence. And every single rejection of scientific realism can be applied to it. Indeed, it's hard not to.

There simply isn't any skepticism or antirealism or relativism that doesn't collapse into denial of physical reality entirely. It always cuts too far and introduces ad hoc backstops whose weaknesses are studiously ignored.[123] The critics of realism have their intense (if confused) moral motivations, and so they grab for rhetorical weapons that cut far more than they have the stomach for or the gumption to pursue.

Surveying the full extent of antirealist arguments leaves you with nothing but embarrassment for the tepid academic provocateurs who prioritized playing to a messy and ultimately dwindling crowd, even when it left them

---

123 Hans Radder, "Empiricism must, but cannot, presuppose real causation," *Journal for General Philosophy of Science.* 52 no. 4, (2021): 597-608.

twisting in knots.

Say what you will about Lamborn Wilson—as a shamelessly inconsistent trustfunder with a passion for child rape and the Shah's slaughter of student protesters—but at least he could trace the implications of antirealist arguments and their asymmetrical skepticism.

OF CHILDREN AND ROCKS

It is no hyperbole to say that *every single scientist of any persuasion* I've ever met has a rant saved up about how we teach science deliberately poorly, how kids are natural-born scientists, and how our education system beats it out of them. Again and again, their self-recognition is not as an elite caste of priests, inducted into a hierarchy, but as *survivors* who, through resistance and luck, managed to keep the flame of childhood alive. This experience of *continuity* was a central contention in the science wars.

Youth liberationists emphasize how adult supremacy is predicated on erasing memory or knowledge of being a child, cleaving our lives and the world into two. This deliberate forgetting, this ritual of epistemic self-mutilation, underpins the domination of children from which every other social hierarchy borrows legitimacy. Childhood is turned into a cipher, a black box, so that children can be turned into objects, pets, accessories, engineering projects. To *remember* our experience would be to meet our younger selves as equals, so instead we construct distant theories about them. We talk about how children are "socialized" in such coarse-grained terms that the actual processes of investigation and the accomplishment of individual insights is removed from sight. All our inductive victories in childhood become smeared into cultural transmission, biological reflex, or the supposed *"a priori."*

Each of these refusals to recognize the child as a scientist is a choice to strip away individual agency.

There is no better takedown of the social constructivist variants of antirealism than just watching a baby examine rocks on their own.

When my best friend's prelinguistic toddler would grab at rocks in a gravel driveway, they were autonomously distinguishing commonalities, patterns, and features within structures that are decidedly unconcerned with humans. Recognition that a vein of quartz makes one rock distinct from the others around it is a convergence not scripted by culture and language, nor is it even strongly conditioned by artifacts in the lens we are handed by evolution. Yes, the visual field involved operates at a certain scale and within certain wavelengths, but there is a strong generality to the staticity, commonality, and distinctness of the quartz vein that *any* mind investigating it would end up recognizing.

> "My experience of the laws of nature in my work as a physicist has the same qualities that in the case of rocks make me say that rocks are real."
>
> Steve Weinberg, *Peace At Last?*, 2001

While those uninvolved in modern science often perceive a sharp distinction between "science" and the theory-crafting about reality we constantly undertake in everyday life—a distinction that centers issues of distributed testimony, powerful institutions, and distant experts—scientists themselves simply cannot make such a distinction. To have done many relevant experiments, to closely know the social incentives and mechanisms, and to be immersed in the evidence and arguments for models, leaves one's experience of science far from abstract.

Sometimes the evidence and arguments are tenuous, in which fields it is only natural that scientists will be more inclined towards hesitancy, pragmatism, and generally low estimations of probability. But in other fields—in particular around the expanse of dynamics we arbitrarily divide up into "mathematics" and "physics"[124]—the evidence and arguments can grow quite strong, overwhelming even.

Thus for many physicists, a skeptical attack on scientific realism wholesale is totally indistinguishable from an attack on object permanence; it's impossible to carve scientific antirealism from metaphysical antirealism in any sustainable way. We simply cannot differentiate the two.

The regularities mapped by Maxwell's equations are not distinctly tenuous *gossip* or *propaganda*, but as tangible, solid, and repeatedly confirmed as our own hands.

Some hypothetical individual who has never had contact with the tug of gravity or the commutativity of adding rocks cannot be expected to assign high credence to reports of such regularities, but if they were to assert their skepticism in sweeping ways approaching binding law, deriding your confidence as foolish, you would naturally rebel and see little difference between such skepticism and a demand that you throw out all object permanence.

Science is a cognitive process of radical inquiry that can be done socially, but also *individually*, because it is precisely what a newborn infant does in learning how to focus their eyes, before any social transmission or collaboration. It's not a method in the sense of a script or formalism—"*come up with an arbitrary random theory somehow and then test it and see if it falsifies; repeat.*" While such nursery rhymes capture some rough outlines of the importance of empirical evidence, they simultaneously put science in a weird straightjacket and evacuate it of pretty much its entire content.

---

[124] Once you discard the extraneous myth of an *a priori* mind that is mystically prior to physics, the relationships mapped by mathematics, logic, and "computer science" are trivially no less empirical qualities of reality than physics, hence the total permeability between them in practice. Nature is under no obligation to respect departments, and constantly makes a mockery of such borders.

> "The scientific method, as far as it is a method, is nothing more than doing one's damnedest with one's mind, no holds barred."
>
> P. W. Bridgman, *Reflections of a Physicist*, 1955

This, I think, points to why physicists were broadly so permissive of Feyerabend. If all you hear is that a fellow physicist is challenging "the scientific method," there's nothing to quibble with. We're not high schoolers in the 1950s who might credulously swallow such tales.

One could just say that the sociologists are defining "science" as a set of modern institutions and cultural norms, whereas the military industrial complex is defining it as an assembly-line proceduralism, and the physicists are defining it as a universally available *orientation* of thought. We could then split the term up into three separate concepts and sue for peace. Each of the three sides could then retreat behind their discursive borders and scoffingly dismiss the others. But this wouldn't resolve the central issue of realism.

We have more pressing adversaries than social constructivists, and any discussion of antirealism that doesn't center the popular ontological variants of the new-agers etc. is irresponsible.

One of the most iconic responses to antirealisms of any sort is to point to a rock and kick it. "*I refute it so!*" This image of rock-kicking or table-pounding is ubiquitous in debates between realism and antirealism. Such arguments can be thought of as doing several things:

1. Retreating to a certain degree of pragmatism in which the utility of antirealism is dismissed.

2. Emphasizing how certain tacit assumptions inherently underlie our interactions. We are, in this sense, revealed as already obliged to realism, and just have to come to awareness of it.

3. Pointing out the immediacy and tangibility of the rock. In this frame, our senses are taken as not just temporarily prior to our conceptualizations, but essentially prior. The "raw" sensation of the rock is implied to be more real than the conceptual model of it.

These arguments are not without some merit, but they are still deeply incomplete and insufficient.

Because, I will argue—expanding on the point of the physicists so often overlooked or ridiculed out of hand in the debate—the distinction between object permanence in the case of a particular rock versus in the case of quantum fields is that there's far *more* compelling evidence for quantum fields.

REDUCTIONISM AS THE BEST CASE FOR REALISM

> "...seeing that the whole then corresponded to its parts with wonderful simplicity, he embraced this new arrangement."
> Galileo Galilei, *Dialogue Concerning the Two Chief World Systems*, 1632

That a rock will kick back a bit when you kick it is still limited and largely unpersuasive evidence for a physical reality; you can stub your toe in a dream. Tactile experience and common sense are simply weak justifications for the existence of physical reality. Even my childhood encounter with the tape recorder is not sufficient to do much more than indicate the existence of *something*.

What makes modern science so compelling is precisely the scope of its successful reductionism; its capacity at *vast lossless* compression of data. Dreams, by contrast, are shaggy dog stories—a host of particulars barely held together by weak and *ad hoc* connections that are not even commutative—at their most lucid they are still constructed only with the resources our brains have on hand.

Science reveals structural regularities and relations giving rise to reality to a degree unparalleled in any other context. The experience of vast complex systems completely reducing to a very small set of dynamics is arresting because it clearly exceeds our widest cognitive capacities.

Learning physics *viscerally demonstrates* the existence of external reality far more than kicking a rock, comparable to how deep and intimate knowledge of a lover's mind can viscerally demonstrate the reality of their existence far beyond the degree to which a discussion about the weather can demonstrate the reality of a stranger at the bus stop. When you briefly glimpse someone out of your peripheral vision or even chat with them about the weather this provides you with very little data, very little depth of modeling their internal structures. They could be a hallucination or shallow simulacra amounting to a chatbot. By contrast, growing to know a person involves being confronted with *novel complexity*, and then their underlying workings slowly revealing itself. Kicking a particular rock once is very weak proof of the existence of matter, the same as chatting about the weather with a stranger is very weak proof of the existence of other minds. It takes wider and more probing engagement.

We are *surprised* by something and then surprised all the more to learn an explanation that completely explains the mesh of relations involved and does so succinctly and in tight interconnection with everything else—akin to the reveal of a tightly plotted murder mystery, but of much grander scope and depth—we gesture at this experience with terms like "elegance" or "beauty" or just run out into the city screaming "eureka!" overwhelmed by the orgasmic bliss of such undeniable contact with the real.

Reality outstrips our cognitive capacities, not in the sense of being unknowable like random noise, but in the sense of having *deep order* behind the initial chaos that we ourselves could not have imagined on our own. This is probably the most philosophically important experience, and one that the physicists in the Science Wars were intimately familiar with. But whenever stated directly, as Weinberg was prone to, it elicited shock and dismissive proclamations of *"weirdness"* from their adversaries.

I think it's a shame this issue was largely dismissed and avoided.

So many antirealists were used to a philosophical tradition where *sensory experience* was taken as grounding, the foundation upon which the rest of science rests. To empiricists, *seeing* or *not seeing* is the ultimate confirmation or falsification. Modern science is then a necessarily more hazy and tenuous project, with "scientific realism" constituting a position over how much to infer from raw sensation. With most academic "antirealists" in this vein declaring some kind of wildly arbitrary cut-off, so that inference to the existence of particular rocks is fine, but inference to the existence of electrons isn't.

But this is ass-backwards.

Not only is sensation often demonstrably mistaken, but we now know that sensation itself is functionally an inductive process, a chain of neural processing that is inseparable from theory creation. Is the electric perturbation from a photoreceptor an experience of contact from light? Well there is always fluctuation, so a certain threshold must be reached for a signal to be *induced*, the fluctuations from multiple photoreceptors are induced together in turn, parsing whether there is an overall intensity or not. Experience is not some distant point or limit on this chain (we could trace the causal flows back to the photons' source or even further) but the very processing itself. That is to say, the process of experience *is* the process of theory creation, any point we might cut between the two would be utterly arbitrary. And one can feel out the shape of this neuroscientific discovery in various ways with a little phenomenological self-reflection.

Additionally, just as there is no realm of pure a priori thought separable from experience, *sensation itself is theoretically laden.* The highest theoretical expectations propagate down and shape the supposedly distinct experiential processing we do. This is not license for "anything goes" mysticism—there are traceable feedback dynamics that press upon and ultimately strongly constrain this whole affair—but it does demolish any reactionary attempt at *just getting back to our foundations, our origins.* There is no such arche that will comfortably shelter us like a father figure. We have no choice but head out into the strange wilderness beyond "pure thought," raw sensation, common sense, and everyday life.

Again, to leverage my own existence as a wild thought experiment for most philosophers: *I was raised in a religion that denies the existence of physical reality*. If you ask me to *default*, my mother's cult did everything possible to set my priors as a toddler to philosophical idealism. The sensory experience of a rock would thus not be as a physical object but as a congealing of thought or belief.

The account of the world as a lucid or shared dream has at least some intuitive pull to a child, and this pull can't be fully broken by anything in "everyday life." The overwhelming asymmetry between dreaming and reality is the *scope of the reductionism possible*. A dream can be, with a little subconscious work, self-consistent, but it has to endlessly pile on complexity because the explanations are being made up as you go rather than solid from the start. Objective physical reality behaves in precisely the opposite way, but the full scope of this asymmetry grows most apparent with a deep knowledge of the contemporary hard sciences.

If I hadn't studied physics at length I wouldn't have been able to fully shake the default idealism of my initial discursive community. Whatever materialism I proclaimed to be more likely, a lingering openness to idealist accounts *would* affect my actions. Assigning a small credence to a model does not mean dismissing that model from all consideration—it has to live with your other models in parallel—nor does it preclude you getting sniped by said model when it can muster severe stakes.

The fractal abuse facilitated and encouraged by antirealism can certainly provide *motivation* to struggle for an alternative, and the staticity of the material in, for example, the case of the tape recorder provides a tantalizing *clue*, but it's the complete asymmetry in compressive capacity of the physical account that provides the most powerful *evidence* to drastically collapse the probability of idealism.

The success of compression, which physicists often shorthand as reductionism, is the strongest evidence that the physical world is real.

> "[Weinberg and company] doubtless have their intuitions, however ineffable, but how different are they likely to be from those that inform Roger Penrose's physicist musings about mathematics, according to which clunky-looking mathematics is a human invention (social construction), whereas elegant, definitive-looking mathematics is a discovery about a realm independent of us (culture-free)?"
>
> Gabriel Stolzenberg, *Kinder, Gentler Science Wars*, 2004

SIMPLICITY, ELEGANCE, COMPRESSION, RADICALISM... REDUCTIONISM

> "The reductionist hypothesis may still be a topic for controversy among philosophers, but among the great majority of active scientists I think it is accepted without question."
>
> P. W. Anderson, *More is Different*, 1972

In 1967, Weinberg noted that a class of particles (leptons) only interact only with the bosons of electromagnetism and the weak force. What, he asked at the start of a very short paper, could be more beautiful and elegant—*natural*—than to unite the bosons of these two fields in the same underlying system of relations, to unite electromagnetism and the weak nuclear force as basically *the same thing*?

Plenty of other physicists had been gripped by the same notion, but the best model at the time still required some mediator particles with masses arbitrarily put in by hand, and it produced infinite terms when describing particle interactions that no one knew how to calculate. Weinberg had been poking around and found (in his terrible notation) a solution that produced those same particles' masses as a result of an elegant "symmetry breaking" early on in the universe leaving behind a trace in the form of a very simple particle field, the Higgs boson.

Over the course of Weinberg's life, he would see his theory confirmed in a variety of ways. Abdus Salam would independently discover the same solution. Gerard 't Hooft would prove that the interactions could actually be calculated. Then experiments at CERN and Fermilab started to indicate the weak neutral current existed, and finally, as technology improved in strength, a string of experiments conclusively proved the existence of each predicted particle from a variety of directions, until consistently measuring them at extremely high precision became trivial.

This electroweak insight would set the stage for later revolutions in physics, revealing other particle dynamics and the early moments after the big bang. Weinberg and his colleagues found it and—before any experimental confirmation—were overwhelmed by certainty it must be true, because the whole of their experience as physicists had instilled in them the conviction that the universe had to be elegant, that it must be possible to compress it in description. In his words, the electroweak theory had, "*a kind of compelling quality... internal consistency and rigidity.*"[125]

---

125 Steve Weinberg, *Dreams Of A Final Theory: The Search for The Fundamental Laws of Nature* (1993), 123.

> "Our job in physics is to see things simply, to understand a great many complicated phenomena in a unified way, in terms of a few simple principles."
>
> Steve Weinberg, *Conceptual Foundations of the Unified Theory of Weak and Electromagnetic Interactions*, 1980

Just before Weinberg's breakthrough, in 1962, Kuhn had published *The Structure Of Scientific Revolutions*, immediately becoming the figurehead of those convinced relativism was an antidote to totalitarianism. But—unlike Feyerabend who would later replace him in this role—Kuhn was deeply perturbed by this fanbase and the conclusions they derived from him. Soon he was admitting there were objective virtues recognized by all scientists in evaluating theories: *empirical accuracy, internal and external consistency, scope, simplicity, and fruitfulness.*

> "First, a theory should be accurate: within its domain, that is, consequences deducible from a theory should be in demonstrated agreement with the results of existing experiments and observations. Second, a theory should be consistent, not only internally or with itself, but also with other currently accepted theories applicable to related aspects of nature. Third, it should have broad scope: in particular, a theory's consequences should extend far beyond the particular observations, laws, or subtheories it was initially designed to explain. Fourth, and closely related, it should be simple, bringing order to phenomena that in its absence would be individually isolated and, as a set, confused. Fifth—a somewhat less standard item, but one of special importance to actual scientific decisions—a theory should be fruitful of new research findings: it should, that is, disclose new phenomena or previously unnoted relationships among those already known."
>
> Thomas S. Kuhn, *Objectivity, Value Judgment, and Theory Choice*, 1972

What's important to note is that all of the above virtues are obviously contained in the notion of *compression*, a pretty straightforward concept in information theory.

Simplicity in the output of a compression requires internal consistency, otherwise the contradictions would potentially lead to infinities. Accuracy is a matter of fit to the accessible wedge of spacetime or phase space, in other words the scope. Predictive capacity is another word for the same when there are (usually temporal) differences in access. External coherence and fruitfulness with other theories is a matter of how much compression it enables on the rest of your theoretical web, and, as we've seen, there's no hard distinction possible between theory and data. Likewise, the value of potential theoretical unifications and reductions fall out clearly.

Yet philosophers—an ideological subculture defined by the privileging of *a priori* flights of thought—have broadly disdained simplicity much as they've disdained induction. And the literature of philosophy of science is littered with declarations that the evaluations of scientists on such must be arbitrary.

> "Simplicity can always be defined for any pair of theories T1 and T2, in such a way that the simplicity of T1 is greater than that of T2."
>
> Imre Lakatos, *History Of Science and its Rational Reconstructions*, 1970

While labelings and pre-given evaluative frameworks can be arbitrarily constructed so as to imply different degrees of simplicity relative to them, the notion that there is no singular objective evaluation of simplicity is just straight up false, and it was *shown* to be false by Ray Solomonoff in his 1964 paper "A Formal Theory of Inductive Inference," with proof independently discovered at roughly the same time by Andrey Kolmogorov in Russia. The critical takeaway being that there is such a thing as *objective* complexity or simplicity, however challenging it can sometimes be to codify or measure. And thus there are objectively more or less effective compressions.

To give a toy example of "compression" for those familiar with basic algebra, consider how the curve that is captured by the very long equation

$$288x^7 - 2256x^6 + 2768x^5 + 19400x^4 - 55750x^3 - 10625x^2 + 175000x - 153125$$

is exactly the same thing as

$$(3x+7)^2(2x-5)^5$$

To switch from the longer description to the shorter description is a *lossless reduction*. We peer into a seemingly chaotic jumble of numbers and extract two simple *roots*, allowing us to capture the curve more succinctly. No content is lost. It would be mistaken to declare that there is only one clearly defined simple procedure by which people can find roots, there are lots of strategies folks can adopt in lots of contexts. *How* you get to the roots is less important than getting there. But there are still better and worse approaches.[126]

A slightly different example, the Einsteinian notion of momentum, is widely referenced in discussions of reductionism in physics. Einstein revealed

---

[126] For space I'm avoiding in-depth discussion of complexity classes. Computing vs. verifying in the case of polynomial roots is very toy and not representative like so many NP-complete problems, but this serves as more of a loose gesture to the dynamic that I think is quickly accessible to any reader who took first year algebra.

that momentum is related to the mass and velocity of an object in a simple non-linear relation:

$$p = \frac{mv}{\sqrt{1 - \frac{v^2}{c^2}}}$$

This can be rewritten into an equivalent infinite sum (a Taylor series):

$$p = mv\left(1 + \frac{1}{2}\left(\frac{v}{c}\right)^2 + \frac{3}{8}\left(\frac{v}{c}\right)^4 + \frac{5}{16}\left(\frac{v}{c}\right)^6 + \cdots\right)$$

Demonstrating that when the velocity is much smaller than the speed of light, all the terms besides the first fall off to very small values, roughly reproducing the p = mv relation of Newtonian physics. Thus the prior model is reduced to a limit or approximation of Einstein's model, mostly accurately compressing much of reality, but only over a strictly smaller domain of it. Einstein's model doesn't just extend in accuracy to a vastly broader extent of velocities and energy scales in which Newton's model fails to apply, and does so with wider predictive capacity, it makes gravity more coherent with other systems, and reveals deeper structural entailments that have consequences in a host of other dynamics. It's successful reductionism, as physicists broadly use the term, successful *compression*.

Even Kuhn realized that the theoretical virtues all morphed into one another from changes of perspective. Desperate to keep his "relativist" flag aloft in some capacity, he treated this as another sign of arbitrariness, akin to the arbitrary evaluation of tradeoffs between Aristotle's moral virtues. But unlike some poorly defined "eudaimonia," we actually know what compression is and that it's an objective fact of the physical universe. The virtues are all aspects of this same thing; just rotations in application, consideration, and labeling.

We would never ask *which is more important: energy or momentum?* Einstein's theory of special regularity (commonly and misleadingly referred to as "relativity") showed us they're different components of the same vector. The only way someone could have a "preference" would be within a very specific context, a specific reference frame, but that creates no inconsistency or arbitrariness, as the underlying universal structure/relation connects them all.

In fact it undermines his attempt to treat them as separate; it's precisely what we would expect in different components of a vector. We can parameterize energy-momentum in different ways, and we will also measure different results in different reference frames, but there is still a single underlying root reality.

In short, instead of *relativism* or *foundationalism,* the science is about the convergence of *radicalism*.[127] We don't get lost in the forest, and we don't respect any shiny marble foundations—we tear shit up and go digging for hidden roots. This is the common orientation between epistemic radicalism and political radicalism.

> "There are a thousand hacking at the branches of evil to one who is striking at the root."
>
> Henry David Thoreau, *Walden*, 1854

Importantly, I am not talking about the supremacy of the micro over the macro *per se*, although certainly radicalism is often a matter of looking under a supposedly "good enough" coarse-grained account of reality for a more finely-grained theory. Radicalism is about finding *maximally successful compressions*, which can mean meta-relations. It is not a simple privileging of the small scale, like some crude notion that only tiny particles are real and universe-spanning laws are not; there's not really a deep distinction between particle, field, relationship, or law.

Philosophers talk of parsimony and whether it's better to have fewer objects or structures in a given theory, but physicists recognize there isn't a difference between structure, object, and relation—they're just different labeling systems of the same underlying realities—and what we're really epistemically seeking is a complete compression over *everything* we have access to, regardless of any arbitrary distinctions between theory and data, reason and empirics. This is why there's demonstrable *convergence* in the theoretical virtues. It makes no sense to evaluate the parsimony of a single theory in isolation.

Of course, as a pragmatic matter we usually do not have the functional capacity to fully evaluate informational compression—just to begin with the universe is open-ended, and it's hard to hold the entire theoretical content science in a single person's head at once. The space of theories, including unconceived alternative theories, may be bounded or at least *pressed* by complexity, but there are still situations where that pressure is not overwhelming. This obliges a certain extent of happy pluralism and diversity as we grasp at different patches of the proverbial elephant, but it is decidedly not an *"anything*

---

127 Philosophers might be tempted to read this focus on lossless compression as a way to systematize coherentism, but the physical realities of the universe applies pressures on the complexity scales of models and this extends quite a bit from the consistency criterion and certainly does not spit out pluralism. Similarly, I wouldn't call compression a truly *a priori* foundation, but rather an inevitably emerging early tendency or pressure on any vantagepoint.

goes" situation. Precise measurements of relative theoretical compression may be impossible, but vast differences can eventually be detected. And since such pluralism is not a moral or social ends *in itself*, but only licensed as an epistemic tool towards the ends of objective truth, there is no justification to keep bad ideas in circulation or demand welfare for dying models. Progress is thus made.

> "Meanings can change, but generally they do so in the direction of an increased richness and precision of definition, so that we do not lose the ability to understand the theories of past periods of normal science... Nowhere have I seen any signs of Kuhn's incommensurability between different paradigms. Our ideas have changed, but we have continued to assess our theories in pretty much the same way: a theory is taken as a success if it is based on simple general principles and does a good job of accounting for experimental data"
> Steve Weinberg, *The Revolution That Didn't Happen*, 1998

Vast success in compression becomes impossible to ignore. Just as the regularity of the universe begins to *force* a child towards the hypothesis of object permanence, so too are scientists *forced* into conclusions as the scope of our access to the universe expands and we discover compressions within it or compressions that unify those compressions.

SCIENTISTS DRAGGED TO REALISM

> "Since [an astronomer] cannot in any way attain to the true causes, he will adopt whatever suppositions enable the motions to be computed correctly from the principles of geometry for the future as well as for the past... For these hypotheses need not be true nor even probable. On the contrary, if they provide a calculus consistent with the observations, that alone is enough."
>
> Andreas Osiander, preface to *On the Revolutions of the Celestial Spheres*, 1543

The history of modern science is, in many ways, a long conflict between those that accepted overwhelming theoretical and experimental evidence as pointing at deeper roots, and those that cautioned against going beyond some fundamentalism or another.

When Copernicus wrote his account of orbits, he certainly believed it was true or at least onto the truth. He was a fervent scientific realist. He could smell the vast *compressive potency* of his model, incomplete as it was, and he deeply associated such reductionist potential with truth.[128] But those involved in the publication were terrified of political ramifications and appended a preface reframing Copernicus' work in an pragmatic and instrumentalist light utterly at odds with his own intent. The role of the scientist was not as a revolutionary, they pleaded and mewled, but as a humble worker in the service of power, merely cranking out *useful predictions*.

In postmodernist-influenced circles it's common to speak of the "*Newtonian mechanistic worldview*," as a derisive shorthand for a paradigm of supposedly reducing reality into simple machine gears to be controlled by capitalists—and Sokal's hoax deliciously cites a number of examples of this phrase—but the joke is that Newton *wasn't* a mechanist. Newton's revolution was in arguing that the universe *didn't* operate in the same way as a clock, with every causal interaction being direct and immediately visible like two gears pressed against one another. His proof of *action at a distance* was received by his contemporaries as mysterious and *anti-mechanistic*, a sign of his scurrilous occult leanings. Digging for deeper hidden truths had a bad reputation; pragmatism was in. To deflect this criticism, Newton dismissed any claim that that the regularities he was modeling were *real*, embracing a somewhat positivistic or instrumentalist approach to natural philosophy, where the only proper aspirations were *prediction* not truth.

Natural philosophers in Britain mostly continued this approach, explicitly rejecting radicalism—speculation about underlying causes—as outside their

---

[128] Indeed his central objection to the ptolemaic model was that it wasn't unified, but arbitrarily patchy

domain, focusing instead on practical measuring and testing. The aristocrats of natural philosophy were more inclined to mapping the *surface* of reality than modeling its roots. They wanted curios to collect and show off, and the Empire needed maps, ledgers, and tables that could be practically leveraged, not speculation of deeper unities. This hostility to radicalism remained so influential that when Darwin gave his account of the species as a product of natural selection he abstracted over any underlying causes.

In many ways, this culture was similar to the pragmatic instrumentalism pushed by the military industrial complex in the mid twentieth century to purge physics of its radicals.

Yet in the 19th century, as participation in science exploded and experimental results proliferated, the tide shifted. Where prior hypothesizing in the Royal Society had been relatively ad hoc and grounded in quite varying epistemic inclinations, the increasingly global practice of science rapidly converged on common epistemic virtues and roughly common weightings of them.

Instead of engaging with subjects and questions in isolation, scientists started to place high value on linking their ideas and conclusions with the ideas and conclusions of other scientists. Consistency and reinforcement in the wider net of evidence and theoretical accounts became more and more important. If a theoretical explanation in one area could also find support from a theoretical explanation in a different area, this was implicitly understood as significant confirmation, perhaps even more so than fitting a curve to your experiment's data. If something to do with rocks and geology could be shown to have a connection with agricultural yields, this reinforced both theories in the crossword puzzle of science as a whole, and the premise that they were substantively speaking about a single reality.

It was an era of unification. And the greatest unification was between electricity and magnetism.

James Clerk Maxwell was a master of, in his words, *"simplification and reduction."*[129] His beautiful account of electromagnetism didn't just unify phenomena as diverse as electrical currents and magnets, but from these principles derived *more accurate* versions of existing theoretical conclusions in optics.

While Maxwell had first, in 1855, spoken of continuous fields from a certain epistemic distance, treating them as merely a useful conceptual tool, as he continued working with fields and tracing their implications he became convinced of their actual *reality*. Granted, Maxwell assumed these were perturbations in a "luminous ether," rather than standalone relations, and Einstein

---

[129] James Clerk Maxwell, "On Faraday's Lines of Force," *Transcriptions of the Cambridge Philosophical Society* 10 (1855): 155-229.

would infamously correct this interpretation. But much of Einstein's accomplishment was to tease out the implications *already* within Maxwell's theory.

As Einstein would describe it, Maxwell's turn to realism, his treatment of the electromagnetic field as *real*, not merely as a tool of *prediction*, was one of the most important developments in the history of science. His account directly rejects antirealist narratives:

> "In the beginning, the field concept was no more than a means of facilitating the understanding of phenomena... Slowly and by a struggle the field concept established for itself a leading place in physics and remained one the most basic concepts. The electromagnetic field is, for the modern physicists, as real as the chair on which he sits. But it would be unjust to consider that... the new theory destroys the achievements of the old. The new theory... allows us to regain our concepts from a higher level. This is true not only for the theories of electric fluids and field, but for all changes in physical theories, however revolutionary they may seem... We can still apply the old theory, whenever facts within the region of its validity are investigated... Creating a new theory is not like destroying an old barn and erecting a skyscraper in its place. It is rather like climbing a mountain, gaining new and wider views, discovering unexpected connections."
>
> Albert Einstein & Leopold Infeld,
> *The Evolution of Physics*, 1938

By 1875 Maxwell would publish "*On the Dynamical Evidence of the Molecular Constitution of Bodies*," arguing for the *real existence* of molecules by knitting together considerations from a variety of different directions, showing how certain strongly confirmed regularities imply the reality of molecules and how the reality of molecules can derive other known regularities. It is a radically realist paper, taking hypothesizing seriously and teasing out implications that explained no existing experimental data nor seemed feasibly testable.

The emergence of such was hardly an isolated affair. Across physics and chemistry, the web of interlocking evidence had grown so vast that it was increasingly realized there was *no choice* but to recognize that certain highly compressive theories really did substantively capture underlying deeper realities. Chemists rightfully remained hesitant about more tenuously extrapolated frameworks like valence, but the reality of atoms and molecules seemed inescapable, as did some relations of composition between them.

In the pragmatist book *Is Water H_2O?*, Hasok Chang traced the history of debates and social maneuverings in the development of our modern concepts of atoms and molecules. Chang wanted to emphasize the social *contingency* of individual moments within the progressive development of the H_2O model, to undermine any objective status of water as H_2O, and so he traced a long story of rhetorical, political, and happenstance dynamics that sometimes privilege an early model against competitors to a degree unwarranted by the evidence then available. Nevertheless, Chang ended up being forced to admit that if some different interpretive branches had been taken, if one of these early victories on the path to discovering water's chemical composition gone to an opposing approach, future evidence and developments would have obliged a correction. There's just so much pressing in from all directions; to stick with certain theories would require spiraling arbitrary complexity in auxiliary hypotheses, abandoning the value of compression.

In a last ditch effort, Chang points out a handful of alternative formulations or approaches to chemically discussing water that could have survived. Water could have been formulated as *one hydrogen and one half oxygen*, by taking the normal gaseous forms of "oxygen" and "hydrogen" (with two pair-bonded nuclei) as their definitional states. Or we could have clung to the old term "phlogiston" and defined it in terms of negative electricity (electrons), expanding our definition of water so that OH- would count as "phlogisticated water." But these are obviously no more than superficial changes of labels over the same captured structural reality. Such minor degrees of freedom in our theoretical notation is not at all a proof that water could be just as easily "*constructed*" as, for example, a compound of silicon and plutonium, a vibrating node in the collective unconscious, a mere collection of constantly churning phenomenological seemings, spiritual projections of water elementals, or a raw platonic solid. While how we have chosen to define "water," involves some fuzziness regarding isotopes, ions, and changing molecular connections, the H_2O structure of three nuclei and the angled binding relationship is a real and vastly general structure. One that would have to be persistent across all theoretical frameworks, in any possible culture, upon sufficient investigation. We are not free to socially construct water in a model with *seven* nuclei, nor does water spring into existence upon us dreaming it, nor do our moods cause nearby water to vibrate and restructure (as was ubiquitously relayed in the '90s).

The space of possible models we could form is not open or all-inclusive, but sharply constrained. So sharply constrained as to have either exactly the same content or almost exactly. So—to turn the parlance of pragmatists back on them—such relabelings would be a *difference that doesn't make a difference*.

The *constraint* of our models can be considered itself a model, and frequently *is* explicitly modeled.

Moreover, pressures of complexity on models, recognized in the virtues of compression, have strong correlation with empirical proof. No matter how much philosophers may protest against the "elegance" that physicists learn to prize, it is precisely this naked embrace of compression by scientists that has driven their empirical success. Those like van Fraassen who would dismiss theoretical virtue for exclusive reference to experimental data and immediate falsifiability—what Samuel Schindler calls a "dictatorship condition"—simply cannot explain the repeated success of science in terms of a whittling down of theories by more empirical data. Theorizing has constantly leaped ahead of evidence, *and against it* when faulty evidence at the time seemed to disprove things, by prioritizing compression.

> "The dictatorship condition has been violated in some of our most stunning scientific discoveries, such as light-bending, the structure of DNA, the periodic table, special relativity, plate tectonics, and the weak neutral current. In all of these cases, the scientists making the discoveries chose to set aside evidence that apparently contradicted their relevant theories and decided to adopt those theories... The virtues gave scientists reasons for belief that their theories were true despite the apparently negative evidence."
> Samuel Schindler, *Theoretical Virtues in Science*, 2018

In recent decades a small coterie of physicists have run a campaign in popular media to denounce string theory, in part on the grounds that—*as was the case for Maxwell's speculations on molecules in 1875*—there is no feasible way to test most of its implications any time soon. In an attitude reminiscent of the logical positivists and continuing the instrumentalism pushed by the military industrial complex, this campaign has popularized among millions the notion that theoretical investigations that don't *immediately* cash out in novel experimental results must be derided and discarded as metaphysical mumbo jumbo. I have no deep stake in string theory itself—I think there's a great deal of valid critiques to be made around institutional funding incentives and overblown rhetoric—but the sweeping derision towards extenuated theoretical exploration as "non-scientific" that has caught on with general audiences during this campaign of know-nothingism is simply reactionary; there's no other word for it.

Such mere empiricism obliges an arbitrary cartesian cut between models and sensation, between theory and data. But just as our thoughts are material processes shaped by constraints and structures of the physical universe, theory itself frequently constitutes data.

Science has never been a mapping of loose surface patterns without consideration of deeper roots.

As the philosopher Richard Dawid has shown—finally giving voice to contemporary theoretical physics in philosophical circles—it's an unavoidable fact that there are "non-empirical" forms of theory confirmation.[130] For

---

130 Richard Dawid, *String Theory and the Scientific Method* (2013).

example, when a theory surprisingly ends up spitting out a known theory or entailing a true theoretical result in another domain, that theoretical result can constitute evidence that your theory is true. Possibility of radical compression on theoretical structures can *constitute a form of proof.*

When, in 1978, two very large seemingly random numbers appeared only one apart in both of the seemingly unrelated domains of finite group theory and galois theory, that instantly persuaded many that there must be a radical connection, the theoretical convergence was too improbable to be mere coincidence. The very smell of a potential reduction in this commonality was enough to assign significant belief that this monstrous moonshine conjecture was true. And it was.

Okay, so scientists overwhelmingly converge on compression as an epistemic virtue that points towards objective structures of actual reality. Why? Why should reality admit such compressions?

BOUNDED COMPLEXITY AND PROBABILITY RENORMALIZATION

> "The fact that induction is (sometimes) successful places a constraint on the world in which we live."
>
> Roy Bhaskar, *A Realist Theory of Science*, 2008

In his "pagan" book, *Libidinal Economy*, Lyotard rejected the framing of structured regularities to reality, an approach pretty close to the Antirealism1 that Lamborn Wilson would gesture at. The picture Lyotard paints is just *event after event*, impossible to capture in any theory, with nothing like an emergent best fit model.

Lyotard would later renounce this book, decrying it as "evil" and "nasty," but he never addressed *why* we should reject such claims, and the book found wide circulation.

As we've seen, antirealists repeatedly find themselves *forced* up the ladder of antirealism into denying all structure and commonality. What starts out as a rhetorical push for humility, pluralism, particularity, or phenomenological fundamentalism inevitably crashes against the universality of modern science, obliging extreme reactions. That physics appears to have discovered lossless (or nearly lossless) compressions in the blinding buzzing confusion around us is a great embarrassment for them. To foreclose the possibility of such universal metastructures, antirealists find themselves obliged to inflate any dynamic of friction, error, perspectivism, or fluidity into a universal absolute.

But—if you postulate the existence of a firm external reality—*the existence of compressions should not be surprising*. Our discovery of and access to them, perhaps, but not their existence. Because for anything, in any sense, to hold together there must be *connections*, which oblige generalizations at some level of meta-analysis. An irreducible universe of full particularity is simply not possible. In order for things to hold together, to have any contact with one another, the overall complexity of reality has to be *bounded* and thus common metastructures in its ontology are inevitable.

Change does not imply discontinuity, it implies commonality along which such change can take place.

Plied with enough drinks, any physicist you meet will end up ranting about how public accounts of the shift between a Newtonian worldview to an Einsteinian one have ignored the revolution wrought by Emmy Noether, who deserves comparable status as the mother of contemporary physics. To summarize crudely, the Noetherian paradigm involves an instinctive familiarity with meta-relations and flows as themselves standard subjects of physics. If how we might gauge something is shown to be arbitrary, we should look deeper at the symmetry relation *revealed* by this arbitrariness to find a solid universal. Her approach proved a connection between symmetries and physical conservation laws, but the *strategies* and *style of thinking* it revealed as useful—in conjunction with the spread of "phase space" modeling—directly undergirds contemporary physics. Weinberg's work on the unification of electromagnetism and the weak force, for example, was solidly a product of this Noetherian paradigm.

While the structuralists were making poetic hay out of *difference*, physicists were becoming experts in mapping *commonality*. In turning to the meta, they sought that which was stable and common. An approach which found great expression in renormalization group theory and generalized into a broader orientation. If each model of reality is a point in a many dimensional configuration space of all possible models, then *we can talk of flows between these models*, with gradual transformations, inclinations, and structure. The story of contemporary mathematics is similar. If you're having trouble with a theory, go meta and work on a *theory of theories* that can provide you with a landscape of them, reveal new possibilities, and allow you to evaluate how likely different regions of theories are.

The only way to avoid the option of conjoining theories into a mega-theory—or a theory of the landscape of theories—is to blindly assert that there's an *absolute* incommensurability between theories. But this is plainly false. No one in history has ever found two *truly* incommensurate models of reality because then we would simply be incapable of any communication or engagement. Moreover they would have to reside in brains in differently structured physical universes, totally out of causal contact.[131]

> "To say that two theories clash or are in competition presupposes that there is something—a domain of real objects or relations, existing and acting independently of their descriptions—over which they clash."
> Roy Bhaskar, *Reclaiming Reality*, 1989

Even *if* some degree of inaccuracy was inherent and inescapable to our theories, if the error of each theory in a set is random (and continuous) then there will almost certainly be a *best* theory. To stop such you would need to contrive the situation so as to have at least some subset of theories be equal in error or remainder, and this meta-structure of equality of probability would

---

[131] One could posit a "mind" in our universe that every step of the way rejects any model with any purchase on the universe, that rejects all empirical contact, theoretical compression, etc.—anything that might lead to overlap and convergence—and thus shares truly no epistemic commonality with Galileo, Bellarmine or even a slime mold, but then it wouldn't qualify in any meaningful sense as a mind.

constitute its own sharp regularity. The realist account of reality would simply be a list of these theories, their degrees and directions of inaccuracy, and their meta-relations.

Because if this space of possible theories can be proven *constrained,* then said epistemic constraint constitutes objective structure, objective knowledge, relating to a concrete reality.

This is important, because there are always those who always seek out whatever scant theoretical wiggle room or uncertainty remains in a situation and act like this proves the entire project of knowledge impossible. They will point to small but continuous variations still possible in theoretical structure or even just potential changes in nomenclature—like our freedom to assign the term "philostogen" to electrons—and try to puff this up into a demolition of all certainty.

But by explicitly theorizing over the space of theories (and the space of theories of theories and so on), this maneuver is revealed as transparently fallacious.

No scientific realist on the planet thinks that our current map of "electromagnetism" constitutes some *complete final best model*; rather we are convinced it captures the root consistencies of reality *to some significant degree*. Vastly more than any known competitor but also in such a way as to constrain *unknown* competitors. We believe that its success at compression leaves open a *sliver* of possible deeper theories, but sharply excludes vaster expanses of possible theories. In short, electromagnetism may be incomplete but it captures content about reality that is unavoidable.

When the antirealist talks of some paradigm shift completely invalidating scientific theories or some noumenal reality that a scientific theory could in no sense capture, it's worth asking this not about some isolated theory but one of the crown jewels of science that is so universally recognized as having all the compressive virtues. *What would it have to mean for electromagnetism to be completely false?*

It couldn't be something like that there exist magnetic monopoles or that quantum field theory has been interpreted a bit incorrectly, because the scope of electromagnetism per se falls short of these possibilities. Developments on such fronts could *expand* our understanding, but they wouldn't mean throwing out electromagnetism.

It couldn't mean merely that electromagnetism is an approximation in a certain limit of deeper, more radical relations. We know that is! Weinberg's unified electroweak field theory corrected electromagnetism with a more radical account, but this did not make electromagnetism completely false. Newtonian gravity is an approximation in a limit of general relativity; and every physicist recognizes the two as sharing much structure, content, and reference,[132] the shift between being pretty much the paradigmatic case of *incremental progress* in truth rather than total revision. Realists *want* a more radical theory of electromagnetism, hunger for the various shakeups to our perspective it might provide, and generally assume such a theory can be found.

It can't even be that there are wacky exceptions studiously hiding that would defy theoretical encapsulation, like random infinitesimal discontinuities or magical demons, because the studiousness by which they would have to hide would constitute legitimization of the structural regularities we *do* find. We could imagine a strange world where mostly just-out-of-sight fairies and Lovecraftian mathematical monsters are crudely *tacked on* to electromagnetism, perhaps only coming out to play at the planck length or somehow outside of anything like our laboratories. We could arbitrarily raise the structural complexity of the universe a good deal higher than expected—filling it with inaccessible fairy realms—but electromagnetism is so broad and reinforced by so vastly many diverse things that our understanding of it would continue to find significant purchase on reality.

We might, of course, be *intentionally and systematically* deluded by something of far greater cognitive resources than we have access to. The classic example is to let every particle in the universe intelligently connive in their exact positions at the dawn of time so as to set in motion an infinitely contrived situation. One might extend the allegorical God who put the dinosaur bones in the earth to test our faith, and say He started a part time gig fucking with magnets and bubble chambers to fool us, but since then we've leveraged His commitment to the bit to the point where we have obliged Him to race around making every transistor in every device on the planet work. In such a case we would be pretty dramatically mistaken about reality, but *not entirely*. Our notion of reference would be fucked and its predictive capacity would be strongly curtailed, but we would nevertheless have correctly mapped a structural regularity, a staticity in the mind of this god (or some framing constraint on the conniving particles). The same applies with any other cosmic conspir-

---

132 Structural realism is popular in philosophy of science these days and is often presented as a kind of retreat from saying "electrons exist!" to saying "the structural relations of the electromagnetic field exist." Similarly, James Ladyman's ontic structural realism argues there is *only* structure. Like many physicists, I am sympathetic but ultimately find distinctions between object and structure or relata and relation to be unhelpful and silly to try and maintain. There is *both* object and structural continuity in science; because one can be turned into the other. The advance from vibrations in the ether to electromagnetic waves is presented as a complete change in objects, yet a continuation of relations. But what is really happening is that there's some content changing between the theories and a lot of content being preserved. How we codify "entities" and "structure" is clearly up to taste.

acy, like Andre Kukla's proposal that reality could be cleaved into physics when we look and but then something wacky whenever we look away.[133]

The only way that electromagnetism could be *completely* false, capable of being blown away in its entirety, is if it was simply not a regularity in any remote way. Something like everything up until this point being complete happenstance, with the bubble chambers displaying something completely different tomorrow, or the shared collective dream having stumbled into a staggeringly consistent storyline for a little bit, before suddenly taking a different turn. Or all our memories could be randomly constructed, like a brain that sprang into being before immediately dissolving. The interweaving of connections between things, all the lines of contact and mutually reinforcing confirmation that we trace when doing the compression that is electromagnetism would have to be completely false. We would have to be mistaken about pretty much *everything*. In short, it would threaten object permanence in a very rapid way.

What is going on here is that the space of possible theories is constrained by complexity.

For any degree of complexity there is a, in some sense, *finite* set of theories that don't exceed it. Such a subspace of theories can be *completely* explored. At some point within the bound there will be *no* unconceived alternatives. Even without a complete articulation, there are meta-theoretical moves of compression that can be made on a finite set of theories to map it and put bounds on the structure and nature of theories. You can increase the scope of complexity allowable in either your theories or reality, but these same core compressions continue, while ad hoc additions and contrivances very rapidly increase obliged complexity. What is simply not possible while retaining anything like thought, intention, and agency is taking ontological complexity to be *infinite*.

And as the ontological complexity to invalidate a well-networked theory is obliged to rapidly rise, the space of comparable theories blows up from very tiny to impossibly large, in which case object permanence breaks down to such an extent that no coherent or directed thought can be sustained in any sort of mind. (And not just physicalist object permanence, but the object-like permanences required by Antirealism3.)

These dynamics are already factored into realist accounts, indeed such meta moves and the objective pressure of complexity and compression constitute the meat of realism. When realists gesture in outrage at the absurd runaway complexity obliged in alternate theories, they are demonstrating the equiva-

---

133 Andre Kukla, "Does Every Theory Have Empirically Equivalent Rivals?," *Erkenntnis* 44 (1996): 137-166.

lence of realism with reduction. What distinguishes the linear sequence-level induction critiqued in the toy models of philosophers from *actually existing* induction is the latter's greedy multidimensional scope and embrace of meta-structural relations.

While many aristocrat natural philosophers of the British Empire were once content with surface-level evaluations of sets disconnected from wider context, and avoidant of speculation regarding universal root dynamics, modern science was inaugurated in the 19th century precisely in the compounding turn away from such relatively disconnected statements like "*all swans are white*" towards more networked compressive efforts. Scientific induction is not extrapolating from one sequence, but over the massive *intersection* of all sequences, including at various levels of meta. All the compressive virtues at once.

Timid antirealist philosophers who want to conjoin skepticism of highly confirmed scientific theories with a noumenal or phenomenal realism *cannot avoid* these dynamics around compression; they're forced to implicitly shuffle such induction under arbitrary categories of things they want to retain as realist grounds, like "seemings."

Similarly, the perpetual attempts by fresh philosophy students to corral "realism" or "antirealism" into definitions that involve arbitrary thresholds of probability estimates is transparently silly. It makes no sense to unilaterally declare some criterion of realism as believing in some discretely isolated theory with greater than even odds (or greater than five sigma odds).

The objective force that realism recognizes is Occam's Razor in the most sweeping sense: the pressures of compression vs. complexity on base rates in the space of theories. How you scale or gauge this confidence in practical considerations is obviously always going to be relative to something else. Assigning 0.000001% credence towards the net truth of electromagnetism against a background complexity of reality where it's more likely all your memories are totally fake is a breathtakingly realist position if it preserves the deeper structures of bounds on theory space.

If you know your cat is inside your house, but haven't figured out what room they're in, you may compare theories of your cat being in the bathroom or your bedroom, but you have an overall deeply realist account: your cat is in one of the rooms!

Towards the end of his life, Weinberg worked on a unification of quantum mechanics and gravity that didn't treat quantum field theory as an approximation of deeper dynamics, but tried to generate gravity entirely within quantum field theory, an approach that was notoriously intractable, obliging new methods of renormalization to wrangle potentially infinite terms down to the finite. When asked, he would freely admit that—gun to his head—he would bet that string theory was more likely to be the true account of reality. But other forms of renormalization had worked in other cases, suggesting they were worth diligently exploring. In a communitarian spirit of science, he was simply happy to do his small part covering all the bases.

Realism *does not preclude* considering multiple models simultaneously, nor does it preclude "humility."

The antirealist, by contrast, cannot ever rise from abject genuflection on their total ignorance; they cannot believe that a theory is even 0.0000000000000001% likely to grasp any fraction of reality. Any ounce of admitted reference or non-infinitesimal probability collapses their entire position. They become bog-standard realists, merely quibbling over credences.

> "To say that [Theory 1] and [Theory 2] are equiprobable is to admit that there is a nonzero probability that [Theory 1] is true, which in turn entails that there is a nonzero probability that the theoretical entities posited by [Theory 1] exist. This is a greater concession to realism than van Fraassen or any other antirealist has been willing to make."
>
> Andre Kukla, *Does Every Theory Have Empirically Equivalent Rivals?*, 1996

The only available move for the antirealist is to claim there is *infinite* runaway ontological complexity, of a kind such that the probability of every sub region in theory space goes to zero or one cannot even begin to evaluate probability at a meta level at all.

It is often noted that the antirealist skepticism promoted by the postmodernists went hand-in-hand with conspiratorial thinking. But there's an important distinction: normal conspiratorial thinking fails to recognize tradeoffs between epistemic virtues as components of the same overall drive for compression. The stalinist is attracted to a simple picture of the world in terms of a saintly communist bloc and a nefarious capitalist bloc, but this obliges immensely complex contortions elsewhere to account for the CIA conspiracy necessary to plant all evidence of Kronstadt or the Hungarian Revolution. At some point the benefits in overall compressive complexity provided by the parsimony in world actors is dramatically overwhelmed by such vast and intricate contrivances. The conspiracist has to *selectively* consider the world, in this case privileging compression at the large scale, ignoring the complexities that such simplifications would make necessary at the smaller scale.

In contrast, while the postmodernists and relativists found great popularity with the selective compression of the conspiratorial or spiritually minded, the sweepingness of their philosophical project forced them into rejecting compression wholesale. Again and again, they were *obliged* to become absolutists, committed to a universe of intractable and utterly irreducible infinite complexity.

The Anti-Reductionists

> "The universe follows very simple laws, and has very simple initial conditions, but in our experience, as classical beings, it's irreducibly complex. The equation on the t-shirt is simple, but the t-shirt is a mess."
>
> Simon Dedeo, *Comedy in Computerland*, 2020

One of the recurring peace offers that the postmodernists and sociologists would make to the physicists (only to be shocked at its rejection) was to accept some level of "*realism*" but then throw out reductionism for some permanent ontological patchiness. The postmodernists were unperturbed by the infinite ontological complexity necessary to stop such patches from simply bridging and unifying into a single underlying system,[134] whereas the physicists were completely horrified.

As we've covered, this pluralistic affinity is rooted in the etiquette of the middle class, where conflicts can be diffused by telling everyone they're right simultaneously and get to remain absolutely right so long as they remain within the scope of their property.

But it's also true that in the daily work of someone like a sociologist, the idea of lossless compression is unthinkable.

Across their myriad texts, Bloor and Barnes spend a ton of time referencing different finicky technical arguments, but for each of their lines of attack, they *always,* in the final lap, had to imperiously dismiss anything like reductionism.

> "Nature will always have to be filtered, simplified, selectively sampled, and cleverly interpreted to bring it within our grasp. It is because complexity must be reduced to relative simplicity that different ways of representing nature are always possible. How we simplify it, how we chose to make approximations and selections, is not dictated by (non-social) nature itself."
>
> David Bloor, *Anti-Latour*, 1999

Now certainly, when engaging with any system in any scale or context, we will attempt to compress it in models that allow us greater traction, and this is often pragmatically motivated. In massively complex aggregate systems like social institutions such compressions involve a loss, they trade away finer accounts, they paper over exceptions, or they neglect entire domains. As a result, in normal life, simplifications are invariably relative to some context,

---

[134] Many philosophical accounts, like Michael Huemer's of "parsimony," make the mistake of judging complexity in terms of component parts. But dualism isn't *twice* the complexity of physicalism, but practically infinitely more complex because it necessitates endless *ad hoc* bridgings or couplings between the particles etc. of the two substances/universes. And there's even further a continuing function of theories of tri-lism, quad-ism, etc. to consider.

vantagepoint, or intent.

Who on the planet would disagree with Rorty when he emphasizes the arbitrarily perspectival nature of many terms in biology?

> "The line between a giraffe and the surrounding air is clear enough if you are a human being interested in hunting for meat. If you are a language-using ant or amoeba, or a space voyager observing us from far above, that line is not so clear, and it is not clear that you would need or have a word for 'giraffe' in your language."
>
> Richard Rorty, *Philosophy and Social Hope*, 1999

While the underlying distribution of atoms is objective, there obviously is no such thing as a *"giraffe"* in the deep ontology of reality, that demarcation or "species" and "organism" is a conceptual simplification that's localized to a certain range of perspective; it does not losslessly compress and is not truly universal.

But the shock of modern science is how starkly *successful* and *lossless* a few compressions are, that is to say how *radical*. We may instinctively look for simplifications for pragmatic reasons, but such strategies work because any interactable physical reality *has* to have connective tissue and thus meta-generalities. These *can't* be infinitely open under interpretation. If we live in a fixed external physical reality then reductionism of some kind is obligatory, and there is no avoiding the fact that physics has been wildly successful at such.

The moment they move beyond challenges to reductionism at the human scale, anti-reductionists are thus forced into moral appeals, grasping for poetic associations in scattershot rhetorical attacks. Latour was deeply representative and, to his credit, frank in this:

> "To put everything into nothing, to deduce everything from almost nothing, to put into hierarchies, to command to obey, to be profound or superior, to collect objects and force them into a tiny space, whether they be subjects, signifiers, classes, Gods, axioms... Tired and weary, suddenly I felt that everything was still left out. 'Nothing can be reduced to anything else, everything may be allied to everything else.' This was like an exorcism that defeated the demons one by one. It was a wintry sky, and very blue... And for the first time in my life I saw things as unreduced and set free."
>
> Bruno Latour, *The Pasteurization of France*, 1984

By grabbing onto the absolute faith that nothing is reducible to anything else Latour reports religious elation, a phenomenological sense of mental release. And this surrender—an Orwellian final victory over one's own epistemic diligence—is equated with freedom.

> "It becomes necessary to admit this first reduction: there is nothing more than trials of weakness... I want to reduce the reductionists... I want to avoid granting them the potency that lets them dominate even in places they have never been... Always summing up, reducing, appropriating, putting in hierarchies, repressing—what kind of life is that? It is suffocating. ... If we choose the principle of irreduction... there is no longer an above and below. Nothing can be placed in a hierarchy. There is no more totality, so nothing is left over. It seems to me that life is better this way."
>
> Bruno Latour, *The Pasteurization of France*, 1984

This sweepingly expansive rejection or "hierarchy" may superficially resemble an antiauthoritarian rejection of *social* hierarchy, but taking "hierarchy" to mean *any sequencing or preferencing* inherently involves rejecting antiauthoritarianism's "hierarchical" ranking of freedom above authoritarianism. In such gestures at "*freedom*" Latour is not endorsing any consistent antiauthoritarianism or anarchism; he *can't*.

In defiance of reductionism, Latour's "actor network" model was happy to slap the term "entity" at every scale and in every possible carving of the universe, describing them as all equally real and impossible to reduce to roots. A table is just as real as a vibe, which is just as real as the electromagnetic field, which is just as real as the commodity form value relation, which is just as real as The World Spirit, which in turn is just as real as arbitrarily contrived concepts like "grue."

This move was echoed by other philosophical novelties like Graham Harman's Object Oriented Ontology, and Harman summarized his and Latour's shared position as:

> "There is no difference between hard kernels of objective reality and wispy fumes of arbitrary social force. Everything that exists must be regarded as an actant."
>
> Graham Harman, *The Importance of Bruno Latour for Philosophy*, 2007

In the immediate aftermath of the Science Wars these approaches found a burst of prestige in the smoking remains of the postmodernist circles. The art world was looking for a fresh new fad with less baggage and a younger generation of continental philosophers were embarrassed by the inane skepticism, credulous pseudoscience, and bad history their teachers had championed. Actor Network and Object Oriented Ontology promised to preserve the pluralist and open-ended interpretive spirit of postmodernism, while sloughing away the more noxious stuff like disbelief in climate change.

And certainly declaring the realness of every conceivable idea, assertion, or relation opens up infinite space to play around in, but the utter sterility of such play is inevitably apparent. Object Oriented Ontology quickly proved no more than an art fad and Actor Network theory has been left at best a rhetorical reference point for a few holdouts in philosophy of science rather than a research programme with traction.

Still, even if the positive alternatives presented by anti-reductionists have petered out, the *critique* of reductionism generated in the Science Wars retains some broad influence in academia.

The philosopher of biology John Dupré was a vicious rhetorical fighter on the side of the relativist sociologists. Like many of them, he denied being an antirealist, but he was openly hostile to the reductionism of physics because he wanted to defend a faith in free will and a set of mystical confusions called the *"hard problem of consciousness."* Dupré saw biology as the quintessential science and was offended and concerned that reductionism would erode the independence of biology from physics.

His strongest point was that biology as a field of research often functionally *cannot* follow the reductionism of physics and is inextricably tied in part to pragmatic human concerns. This is certainly very true!

> "Apart from its lack of commitment to essentialism, the degree of naturalness of kinds will depend on the extent to which one thinks of scientific theory as independent of human interests, a question that might be answered quite differently from one area of science to the next. If any scientific inquiry is conceived of as addressing questions the significance of which is partly determined by particular human interests or needs, then the best possible theory in this domain will delineate only natural kinds relative to the interests in question. This is not antirealism. Nature may well determine what is the best theoretical approach to these problems. But other problems would imply other theories and other, equally natural, kinds."
>
> John Dupré, *The Disorder of Things*, 1993

But the radical realists agree with this much! We've never conceded that all of biology is as radical as physics, we've always noted that many arenas of biological investigation are forced into compromises and arbitrary or pragmatic conceptual choices.

In *Science as Radicalism*,[135] I traced some of the history of the nineteenth century appropriation of the mantle of "science" from physics by early doctors holding very different orientations, practices, goals, and inclinations. This is still a relatively fresh theft! And physicists have remained tetchy about this wider (mis)use of the term "science"—in particular the unwarranted confidence it instills in other fields while they deviate from radical analyses. Physi-

---

135 William Gillis, "Science as Radicalism," *Human Iterations,* August 18, 2015, https://humaniterations.net/2015/08/18/science-as-radicalism/.

cists are notorious for grousing about the situation in terms like everything is *"either physics or stamp-collecting."* [136]

This does not mean biology is not *empirical* or that it is without scientific content, practice, or influence. Far from it. Regularities clearly exist in biological subjects. When we talk about DNA we are clearly talking about a broad fact of nature, however hazily gestured at. Indeed *parts* of biology plumb quite deep towards intricately truthful reductionist accounts. But biology is systematically plagued by the roughnesses and fuzzinesses obliged by its scale of abstraction in description.

Because of this unavoidable haziness in the slicing of labels used by biologists, it is often not possible to cleanly define a biology-scale category (like a "species" of bird) in terms of some symmetry in the underlying physics. But it's completely absurd to then conclude not that biology is epistemically constrained by pragmatism but that it represents a patch of ontological objects and forces independent from physics, as described in Antirealism4.

Dupré repeatedly gestured at something akin to Antirealism4, only to at one point relent in his rhetoric and grudgingly and dismissively admit that *nature itself* might determine the one true best approach. One can only imagine physicists and philosophers dedicated to foundational questions of reality screaming, *"Yeah, uncovering that is our whole fucking job!"* But Dupré can't bring himself to care about such, nor can he really imagine an audience who does, because he's interested in *"other problems"* and other *"particular interests or needs."*

And it's not universally morally wrong to adopt pragmatic or human-centered conceptual lenses. These are important! Every single one of us uses conceptual shorthand in daily life that is partially shaped by our pragmatic needs and aspirations rather than cutting reality at universally objective joints.

The radical realist position is not to immediately establish every single claim at the abstracted layer of biology in precisely reducible terms and reject everything else. A doctor is not a scientist but more akin to a technologist, proactively bridging and negotiating between the domain of natural fact and the domain of human intent and need. He may use scientific approaches plumbing for deepest root causes and universals, yet he will often choose the merely "good enough."

By contrast to such pragmatism, the dynamite-chucking *radical* knows that while life is filled with imperfections and can oblige some temporary compromises, you should never *settle* for "good enough." Struggle has left her intimately and sharply aware of how dangerous "good enough" can be, how a compromise in pursuit of immediate benefits can limit her horizons and create

---

136 J.D. Bernal, *The Social Function of Science* (1939), 9.

obstacles to further progress. She knows that to coarse-grain the world into nations or institutions and treat them as *actually existing entities* of independent or even *greater* firmness than the individuals they are composed of is to blind oneself both to oppression and opportunities for resistance.

The coarse-grained theory may have an effective domain of description, some patch on a map with a border, but it is the purpose of the radical to search beyond, for anything that could shatter, explode, dissolve, or otherwise *reduce* the terms of that theory, even if it means a more complicated picture. This is what she means when she declares that individual human beings are *more real* than "nations."

> "What is a nation? Show me one! I don't want to be killed by an abstraction."
> R, *Death by Hanging*, 1968

> "It is the symmetries of the lower level that dictate the allowed interaction terms at the higher level, but not the other way around."
> Thomas Luu and Ulf-G Meissner,
> *On the Topic of Emergence from an Effective Field Theory Perspective*, 2019

8

REALISM AND LIBERATION

"I'll admit that you were right about the potential for science studies to go horribly wrong and give fuel to deeply ignorant and/or reactionary people."
Michael Berube, *The Science Wars Redux*, 2011

"The rise of postmodernism has totally discredited the necessity of, and even the possibility of, questioning the inherited metaphysical systems, which for centuries have shackled human imagination and social freedoms in those parts of the world which have not yet had their modern-day enlightenments."
Meera Nanda, *Prophets Facing Backward*, 2004

"Anything is okay in a good cause."
Barry Barnes, *Rationalisation by Experts and their Regulators*, 2014

My mother has just committed a cruel and vicious act, laughing in glee while doing so, perhaps painfully twisting my tiny fingers in icy water, pushing my sister out of a car on the side of the highway, throwing our tiny bodies around the house, or holding me prone while forcing some noxious food down my throat while I convulse on vomit.

A few seconds pass.

Suddenly the new crime is that I'm silently sobbing. A harsh interrogation starts in which it is a mistake to admit what just happened indeed happened. Any recollection or assertion of objective reality, even if observed silently the whole time by sibling bystanders, brings furious recrimination. *"Does it make you happy to believe your own mother would do such a thing?! Are you happy as a result of your belief?! Why won't you just accept the reality that will make you happy?!"* she half snarls, half screams.

Why can't you just be pragmatic?

It's time to turn more fully back to the stakes.

THE MUTUAL ENTAILMENT OF COMPRESSION WITH AGENCY

We've seen that a completely pragmatic and instrumental orientation towards epistemology directly opens you up to exploitation.

Folks can always contrive to create a situation, both social and material, in which it *benefits* you to believe not what is true but whatever they want. Once this crack is opened it can spiral into a feedbacking dynamic until your entire mind is arbitrarily controlled.

But, even without fixing a specific belief in your head, someone can control you simply by cutting off your ability to recognize or approach the truth, locking you in a passive humble stasis, curtailing your ability to predict the results of your actions.

Even if you don't end up asserting that there are five lights instead of four, the agony of torture has neurological effects that diminish your capacity to recall memories accurately or evaluate evidence rationally.

In countless ways, power studiously works to produce *uncertainty*.

The most common response of postmodernists is to retort that you don't need to believe in an objective truth to resist such control, all that's required is *autonomy*. Just assert your distinctiveness, your *difference*, and defend it to the death. Belief in astrology or spiritualism or that you can will money into being by focusing on a sigil, are all as good as any other belief—indeed the more different from the status quo, the better! If you *really* want to be ungovernable, we are told, then forget organizing insurgent strikes on oil pipelines and instead addle your brain with psychedelics and annoy the local Food Not Bombs with your tales of machine elves.

Suffice to say, in the struggle against power, such embrace of "difference" hasn't made a difference.

I discussed earlier how legibility isn't the same thing as domination and illegibility isn't the same thing as freedom, but that was a surface examination in the context of social organization; covering how illegibility can facilitate social hierarchies and legibility can erode them. It's worth emphasizing as well that underneath the political call for *social diversity* as resistive to power is always a moral attachment to "difference" as *individual autonomy* in a very weird sense of "autonomous."

Just as anarchists did, many philosophers immediately recognized this as a conceptual inversion of "freedom."

> "In-gear freedom is a matter of interacting causally with the world in order to realize our intentions; it is threatened by any view which denies the efficacy of our intentions in bringing about changes in the real world; out-of-gear freedom is precisely a matter of disengaging our choices from causal interaction

> with the world, to ward off the threat that the nature of that world might limit or determine them. One instance of an out-of-gear conception of freedom is expressed by Rorty: 'Man is always free to choose new descriptions.'"
>
> Andrew Collier, *Critical Realism*, 1994

That is to say, many people have implicitly taken freedom as a matter of being at least partially *un-caused*, a patch of reality somehow severing its inputs from everything else.

The argument is basically: *if someone or something caused you to do something, like arrive at a specific description of the world, are you really free?* Thus the embrace of supposed randomness and chaos by figures like Lamborn Wilson as well as the continuing popularity of vitalism on the Left in various guises ("wildness", etc.). In this frame—which Foucault's terribly broad use of "power" helped reinforce—for us to speak of power being defeated, not just rationality but *causation* must be defeated.

Such notions of freedom as avoidance of external influence are incoherent and impossible, since no one is a cartesian island, able to disconnect from the world and return to some primordial *"a priori"* authentic self.

The drive to do so is fundamentally reactionary, and it necessitates slicing up the universe into patches and fiefdoms, nations and discrete identities. In contrast, anarchists know that it is not our *barriers* to the world but our available *connections* with it that are what instantiate us and give us freedom.

What's being leveraged in the sentence *"if someone or something caused you to do something, are you really free?"* is not the very existence of any causal connection but the implicit slicing away of *other* lines of causal connection, other possibilities.

Just as nationalists try to violently slice apart the entangled web of humanity, domination more generally is about trying to slice away your connections with the world. Shackles and prison bars aren't so much a matter of being overly connected to the world as they are a means of cutting off all your *other* possible connections to the world, shrinking down your causal cone.

In terms of information, this is why if someone isn't sufficiently or accurately informed we recognize that as a violation of their agency. If you lie to someone you take home about your STI status, you undermine their agency around choosing to sleep with you.

Freedom is a matter of choice and choice is not some magical cipher beyond understanding but a *deliberative process* in the material system of our brains where we take in information from the surrounding world and evaluate it. Not from some distant magical remove, but through compressing that information into models so that we can predict the consequences of our potential actions—the connections extending out from the evaluative process into the future. Choice is thus a dynamic of *compression*, whereby causal connections are knitted efficiently, allowing the small to impact the big. Enabling us to do more than passively "*resist*," reactively following an immediate gradient, but to leap beyond, to tunnel into possibilities otherwise inaccessible.

And an incompressible universe would be a universe without agency. In such "ontological chaos" choice would be impossible, incoherent.

While claiming that radicalism—i.e. reductionism—must be possible, I made two arguments. The first was an ontological argument that any accessible reality must have common connections which enable global compression, just to be able to hang together.

The second, however, was an implicitly transcendental or anthropic argument that there is a related complexity bound on the conditions of thought and agential or conscious existence itself. This is distinct from the pragmatist argument that we simply find compression *useful* towards specific ends, because it strikes at the necessary universal premise of cognition, prior to any value set or cultural context.

Because the level of ontological complexity necessary for some alternative theories blows past this bound, we can discard such theories. It "could" be the case that all your memories and sensations at this very moment are randomly constructed and you will cease to exist in another moment, but at such contrived complexity scales, utterly resistant to lossless compression, there simply is no choice or agency. You would not be able to evaluate your inputs in any way that would grant traction on the universe. The process of informed choice would simply not be possible. Such skepticism thus erodes not just "external" object permanence but internal object permanence and all possibility for thought.

And we can fence in the space of potential alternatives to any scientific theory by teasing out how rapidly they will have to grow in complexity.

It may make you *happy* to believe that your owner has never whipped you, it may buttress some fortress-like inability to be reached, it may serve any number of arbitrary goals, but to disbelieve the overwhelming compression of *it happened* from all your memories, physical wounds, etc. is to abdicate thought. There is a deep sense in which the very condition of your existence as an agent is predicated upon seeking the truth. In a similar sense to how you cannot gulag a population into being free, you cannot delude yourself into being free.

Agential freedom and radical truth-seeking are *constitutively* intertwined as ends.

It's not merely that knowledge gives us technology by which to better move the world—because sometimes truth does not end up spitting out useful tools and a crude engineer's pragmatism would occasionally license self-deception and broader enslavement—it's that the pursuit of truth is simply part of *what it means* to be agential.

There is thus space to elevate the pursuit of objective truth from a pragmatic goal of some utility towards other ends, into *an end in itself*. This provides the agency of accurate maps and the resistive grounding against aggressive manipulation.

Contrary to Rorty, reality is not a master that realists prostrate ourselves before in hopes He'll be more gentle, but—if it must be personified—a co-conspirator who provides us the nourishment to keep our minds alive.

The View From Anywhere

While many liberals whined that a single reality would be "totalitarian" and "monotheistic," more frequent were sneers that realism constitutes "a view from nowhere."

> "The whole content of realism lies in the claim that it makes sense to think of a God's Eye View (or better, a view from nowhere)."
> Hillary Putnam, *Realism with a Human Face*, 1990

One response is to object that a single physical reality is not the same as a single correct view or *relation* to reality. It doesn't particularly matter if you describe water as "H_2O" in one language and something else in a different one, so long as the same structures of reality are being compressed and referred to. Every society on this planet probably has a term for the Sun, but that doesn't mean they perceive different Suns. Similarly, an alien species will be constrained by the same physics as us, they may encapsulate those relations slightly differently, but they'll face the same objective pressures on how to compress them. Learning different languages or perspectives and translating between them is always possible when they refer to the same reality.

What I think is politically revealing, however, is how the sneer of the "view from nowhere" privileges the reactionary fetish of placeness.

Like a person who strides up to you and demands to know "where are you from?!" this kind of thinking demands that every mind be stamped and catalogued, that everyone *has a place*.

For what does "the view from nowhere" connote but the perspective of the refugee, migrant, itinerant, and nomad. An individual without a consistent or fixed placeness, without the kind of fixed ties to the land, culture, and history that Heidegger lauded, could well be said to represent the "view from nowhere." Indeed the eternal fascist disparagement of "globalists" and "cosmopolitans" has always explicitly been grounded in fear of the dangerously infectious universalities gravitated to by those without a home, those untethered and mobile. To be stateless is to be untrustworthy, and to *choose* to be stateless is to be a traitor. Opposition to universalism always comes tied at the hip to opposition to movement.

When I was homeless as a young child, I clung to the few possessions I had left: an action figure and a dog-eared copy of *Jurassic Park*. Both were lost when my sister and I were abducted, held captive, and almost sold by a woman at our shelter. From that point there was nothing but flux. An entire day would be spent sitting in one or another welfare office with my mother. The next day scrunched up against a concrete wall outside watching people walk by. A night in a dingy motel room provided as a charity, greedily devouring a can of green beans. You learn to speak so many different languages in wildly different contexts, to play different games. To be different people.

And, deprived of any fixed points, you begin to see the symmetries, the commonalities that apply everywhere. You have to remake yourself so many times that you prize the insights you can carry into *any context*, the underlying dynamics that express themselves differently but have the same bounds or rules. Whatever you were originally becomes sloughed away under the endless transformations.

This emergent universalism is the nightmare of fascists. All around the world, in every culture, bright kids leave their small towns and come back irreversibly transformed. No matter where they originate from or where they travel, they end up convinced of many of the same alien beliefs. Weird ideas that spontaneously arise and flourish the more their children travel, the more contexts they are thrust into. To the parochial reactionary watching these transformations, this effect is a lovecraftian horror.

Their children return not as foreigners but something more terrifying. With no deference or fealty to any local perspective, they become the children of nowhere, of everywhere, of anywhere.

> **Reporter:** *There are some people here, roaming about... well not exactly roaming, they seem organized. I don't know who they are, they're all dressed in black, they have black hoods on, and black flags... a flag with nothing on it.*
> **Anchor:** *A flag with nothing on it?*
> **Reporter:** *That's right, it's totally black.*
> News Report, Seattle, November 30th, 1999 [137]

Before the air became saturated with tear gas and smoke, it had been impossible to miss the contradiction in our signs and chants. The seventy thousand protesters who converged on the streets of Seattle from around the world may have been united against the WTO, but for very different reasons. Some of us identified as "anti-globalization," others as "alter-globalization." Many denounced the emerging global order as a matter of unequal freedom, where capital could cross borders but people couldn't, creating de facto prison camps in the global south for sweatshop labor. But others denounced the very idea of any global culture whatsoever, making no distinction between McDonald's and feminism, in language very similar to Foucault's praise for the Iranian Mullahs as defenders of local particularity.

---

[137] Richard J.F. Day, *Gramsci is Dead: Anarchist Currents in the Newest Social Movements* (2005), 1.

Even as a young teenager, I couldn't help but feel perturbed by the implicit tension dividing up the masses in the streets.

Were we advocating an alternative globalization from below or were we micro-nationalists opposed to movement, connection, and consensus?

Were we about *solidarity* or about *difference*?

In the years that would follow—as anarchists would painstakingly expel fascist entryists from ranks of the Left, and as fascists would incessantly whine that anarchists had somehow *betrayed* the the anti-establishment coalition of the '90s to side with the "globalism" of Nike and Starbucks—this tension was brought into ever sharper relief.

Anarchism is predicated on the notion that your freedom and my freedom do not have to conflict, but can instead conjoin and compound. In contrast, fascists view "freedom"—when they acknowledge its possibility at all—as a zone of property, a zero-sum game, assuming that everyone's values are inherently particularized to the arbitrary context of their birth, individual against individual, family against family, tribe against tribe, nation against nation. Thus for some to achieve freedom, others must lose theirs.

Realism dissolves this conflict by making freedom concrete and material rather than subjective and affective, a potentially *universal* measure across the unruly web of humanity, where the accessible space of possibility can increase for everyone. And it enables objective convergence on values; because the extent of anyone's choice is dependent upon accurate maps of reality, collectively improving our maps increases *everyone's* freedom.

> "The freedom of all is essential to my freedom. I am truly free only when all human beings, men and women, are equally free. The freedom of other men, far from negating or limiting my freedom, is, on the contrary, its necessary premise and confirmation."
>
> Mikhail Bakunin, *Man, Society, and Freedom*, 1871

Just as anarchism pushes for extension of our circle of care to *every* agent, not just to the stranger or foreigner but to children and animals, realism is likewise grounded in an empathetic move towards considering *every possible perspective*.

What the fascist derides as impossible—as a view from nowhere—is the view *approaching everywhere*, the view *from anywhere*. The common firmnesses still found by those in motion.

The swing of the pendulum and the motion of the planets seem to occupy two very different domains, but the traveler between them can find the same unchanging underlying root.

REALIST USES OF STANDPOINT AND DIVERSITY

In *Science as Radicalism* I emphasized science as an *orientation* and *desire*.[138] This was explicitly to emphasize the utility of alternative lenses on scientists as something closer to an ideological movement, minority identity, or even sexuality. In so much of the Left's discourse, "science" is treated in terms of institutions or grand philosophical topics, and what's evacuated from that is any obligation to empathy with science as a perspective in the world by actual human beings.

The term "perspective" often carries relativist associations, but it need not. Sometimes a single perspective is just the fucking correct one. Anarchism is a *perspective*, and it is also—trivially—*objectively right* to a degree that can license violence. Those seeking the universal abolition of slavery were at many times in history a fringe minority, they were also absolutely and objectively correct. I firmly believe that scientists could stand to build a little more fire in their bellies and to cultivate self-perception as a tiny minority of righteous insurgents in a world that does not share their values or goals. By definition, every scientist is a *radical*, but unfortunately this often goes without a militant spark.

Similarly, once you define science as a *perspective* rather than some institutional allegiance, it becomes incumbent to ask what unique advantages and insights that perspective might have.

What might scientists contribute, not as an obedient demographic in some grand Leftist coalition, assigned to periodically turn in reports on climate change, but in a richer and more aggressively challenging sense? In their situatedness, scientists engage with deep aspects of reality other people never see, and build up not just epistemic tools but intuitions and tacit knowledge. The *lived experience* of a scientist's grasp beyond everyday life enables a more fertile vocabulary of metaphors, an awareness of twists and turns possible in conceptualizing.

This is an appropriation or adaptation of arguments for respecting and elevating the perspectives of oppressed minorities, but I do not mean it ironically or with hostility. Certainly not in a way dismissive of the value of deep engagement with the insights of e.g. queer or indigenous perspectives. I broadly agree with such appeals.

I did not discuss "standpoint epistemology" or "viewpoint diversity" in my initial summary of the stakes to realism and antirealism, despite these topics regularly coming up in public fights over "postmodernism." The reason is

---

138 William Gillis, "Science as Radicalism," *Human Iterations,* August 18, 2015, https://humaniterations.net/2015/08/18/science-as-radicalism/.

simple: arguments for both *presume realism.*

And there is simply no question that they are useful strategies and heuristics in the service of grasping reality better.

> "Suppose we had a caste system that sent 50% of people to factory floors every day and 50% to office buildings, and never the twain shall mix or visit the other's place of work… An empiricist informed of this arrangement should immediately conclude that the blue collars are much much more likely to know about factory floors and what they are like when compared with white collars, and vice versa for office blocks… the marginalised are clearly in a better position to know what's going on and ask pertinent questions."
> Liam Bright, *Empiricism is a Standpoint Epistemology*, 2018

Standpoint epistemology is fine, within reason.

It's trivially true that the oppressed often have more relevant and piercing knowledge of their conditions and the structures or mechanisms of their oppression. A common coping tactic of the abused is to develop more intense faculties of empathy and modeling, since more is on the line for them, whereas the beneficiaries of oppression are more often vacuous, uninquisitive, ignorant, and slow. Poor kids are generally far smarter than rich kids, because we have to be. Women have often put great thought into mapping the internal lives of men, who themselves rarely do the same. An ethnic minority may become adept in multiple languages, certainly multiple codes, and between them feel out more of the underlying proverbial elephant. A prisoner becomes an expert in the psychology of their torturer. The privileged simply aren't *alive* to the same degree as the rest of us, and they frequently resent us for this, becoming fixated on authenticity, intensity, and risk in clumsy attempts to patch the lack in their lives. They resent the wound that their power over us inflicts on them, and will never forgive us for it.

There are exceptions. Access to some knowledge and experience can be denied to the oppressed and hoarded by the oppressor. At different levels of society different styles of power games with different rules may dominate. The oppressed become experts within the domains they have contact with; they can't be expected to be sommeliers. At the same time, the privileged often dramatically overestimate the extent of the knowledge domains they have captured.

Viewpoint diversity is fine, within reason.

It's trivially true that theories take time—sometimes centuries—to sort themselves out, to develop sufficient resources to challenge, tweak, or defeat existing theories. While there is an objectively limiting fact of compression, we usually do not have the resources to judge variations in it to great precision; when broken apart into epistemic virtues the imprecision multiplies. And much as we are limited in our capacity to view the future, we are often quite limited in our capacity to evaluate the ultimate fertility of a research program, which means that long tails of perturbative variation can be necessary in a society to feel out opportunities.

There are exceptions. Theories must still be winnowed. Massive differences in overall compressive capacity are often quite clear, as are severely skewed proportions of the compressive virtues. And when it comes to ethics, there is no need to keep open the door forever to the possibility that genocide is good, nor is there social utility in providing the welfare of a platform to genocidaires. Failing to curbstomp evil values is simply complicity in evil. You retain a moral obligation to lure nazis into the French countryside and garrote them, to plant bombs in the offices of local gendarmerie, and to execute snitches, duh.

Standpoint epistemology and viewpoint diversity are *instrumental heuristics*; they are not universal context-independent obligations and can of course be misapplied.

Such misapplications have been, of course, repeatedly embraced by the waves of "identity-politics" obsessed liberals of the last two decades, who consistently steal the basic insights of radicals, crudely regurgitating them in ludicrous ways designed solely to further their class incentives.

> **Worker:** *They're calling the Black and brown workers behind this [unionization] colonizers. And getting white men like you to spread this message. Disgusting.*[139]
> **Union Buster:** *False*[140]
> **Worker:** *What is false? I work there lmao*[141]
> **Union Buster:** *Aren't there always multiple perspectives?*[142]
> Exchange on Bluesky, 2025

> "In my experience, when people say they need to 'listen to the most affected,' it isn't because they intend to set up Skype calls to refugee camps or to collaborate with houseless people. Instead, it has more often meant handing conversational authority and attentional goods to those who most snugly fit into the social categories associated with these ills—regardless of what they actually do or do not know, or what they have or have not personally experienced."
> <div align="right">Olúfẹ́mi Táíwò, <i>Being-in-the-Room Privilege: Elite Capture and Epistemic Deference</i>, 2020</div>

In my mind, these are so obvious that it wastes everyone's time to bring them up. They may be pressing matters for inane liberals/conservatives fighting for status in nonprofits or journalist groupchats, but every anarchist or

---

139 https://archive.is/Vt3Xc.
140 https://archive.is/A32Bj.
141 https://archive.is/d76wq.
142 https://archive.is/OCDKk.

leftist I've ever met recognized these dynamics immediately.

Of course, this reveals how a certain percentage of the bluster and posturing around postmodernism derives from one's social position in such discourses, and wheter one views the political landscape as two monolithic buckets of liberal versus conservative.

Those in closer contact with liberals/conservatives are more likely to see value in rhetorical bravado to emphasize the importance of standpoint and diversity, if only as a crude rhetorical cudgel by which inane liberals can push back against inane conserativces. But those of us in zero contact with liberals/conservatives—those, for example, born and raised within the radical Left—can't help but look on in horror.

Whatever *discursive* or *pedagogic* utility there might be in ignoring the exceptions when one is part of a social dialectic with grunting troglodytes—there is no *object-level* utility to ignoring reality among one's own.

One of the most tell-tale determining factors in the gestation of authoritarian communists in the US is how isolated they were from the Left or radical subcultures.

When you listen to the life stories of rank-and-file american stalinists what stands out is how often they had their initial political development *alone*, amid a sea of liberals/conservatives, without even a single other leftist in sight. They never even met an anarchist until it was way too late. And they were never in a space or involved in a struggle, conflict, or project where they could imagine a concern other than fighting the status quo. Later they would inevitably be snatched up by one or another cult. But at the start, all their reference points were The Enemy, and their ideological development was just pragmatically grabbing at any argument that pointed at The Enemy.

When you begin to learn how badly the state has lied to you, how completely gullible and culpable everyone you know is, there is no choice but to look at every newspaper and academic book as CIA propaganda. It makes a kind of brute sense, in such a world, to not trust anything. To decide to commit leaps of faith, casting off into the void.

If the Great Satan says that X is bad, well maybe that should count as evidence that it is good. Even if X is bad, *who cares*, by fighting for it at least you will go out doing some good resisting the Great Satan.

If it's impossible to find truth and all you have is pragmatism, well then the important thing, the stalinist resolves, is to commit to the bit.

Let future generations sort out any mistakes.

9

THE REACTIONARY ROT IN POSTMODERNISM

"Entire Ph.D. programs are still running to make sure that good American kids are learning the hard way that facts are made up, that there is no such thing as natural, unmediated, unbiased access to truth, that we are always prisoners of language, that we always speak from a particular standpoint, and so on, while dangerous extremists are using the very same argument of social construction to destroy hardwon evidence that could save our lives."
                                            Bruno Latour, *Why has Critique Run Out of Steam?*, 2004

"Idealism—it doesn't matter what postmodern people say—is an inherently conservative philosophy."
                          Manuel DeLanda, *Columbia Art and Technology Lecture*, 2004

"The problem… was not what our enemies did but what our friends did."
                                    Hannah Arendt, *What Remains? The Language Remains*, 1964

There is, of course, a very trivial sense in which postmodernism is by definition a continuation and deepening of the reactionary project. Since the French Revolution, "radicalism" has stood for the project of seeking to better grasp reality at the roots and with that knowledge being able to Change Everything. The project of "reaction," expressed most influentially by Edmund Burke, is an outraged endorsement of humility in response.[144]

As we have seen, postmodernism is, in almost every way, an attack upon radicalism. It is not a distinct or original project, but a direct continuation of the reactionary legacy, obfuscated and rebranded to appropriate a veneer of progress, but really just the same old thing.

Nietzsche and Heidegger's children did not stray far from the nest. The academic Left reacted to the capture of the hard sciences by the state—as well as its erosion of their radical orientation—not by doubling down on radicalism or assisting insurgent scientists, but by getting into bed with lingering aristocratic reactionaries who were even worse than the liberals that had displaced them.

In an utter historical grotesquery, the starkly particularist and anti-universalist values of the nazis, were adopted and propagated under the guise of critiquing said nazis.

The Left did this for incredibly banal reasons:

1. **Shortsighted preoccupation with the near-enemies of whatever the current struggle was.** This immediatism came packaged with a rejection of striving for the future. Gone was anything like the gumption that had motivated far-sighted anarchist prediction of how marxism would turn to tyranny. Concern with potential future enemies (or resurgent past enemies) was ridiculed as a distraction and disconnected intellectualism.

2. **The lingering influence of democratic populism, mass-fetishism, and holism.** They instinctively viewed the enemy-of-my-enemy as a valuable member of a big tent coalition. If the only enemy was a single mass homogeneous bloc, the only response they could imagine was to

---

144 The modern anarchist and feminist movements have their roots in the duo of William Godwin and Mary Wolstonecraft, the latter of which wrote the famous essay A Vindication of the Rights of Men in outrage at Burke's defense of epistemic timidity and endorsement of tradition on such grounds.

create something similar as a counter. Thus all deviants or enemies of the mainstream must be embraced in a single Party, with all the abhorrent "tolerant pluralism" that entails.

3. **Reactive cultural parochialism, whereby even mere aesthetic associations of the singular enemy had to be purged.** As the hard sciences were conquered by the state, the overall compounding reaction by the Left was to cut them loose, to demonize or ostracize as unclean anything that touched their circles. A strong current came to instinctively view the Left as definitionally that-which-was-not-science.

None of these are new mistakes within the Left, but long standing failure modes stemming from its rejection of anarchism and methodological individualism. This saga of catalyzing hostility towards science is but one facet of the Left's overall rejection of antifascism.

While it may come as a shock to younger generations, antifascism was relatively marginalized and pretty openly despised within the Left in the '80s, '90s, and '00s for many of the same reasons science was. When (primarily anarchist) antifa activists coalesced in response to fascist street violence, and fought back, including shutting down Holocaust denier talks, state communists immediately recognized that a prohibition on genocide denial posed an existential threat to them. Their endless little newspapers convulsed with horror at the spectre of antifascism, regularly denouncing the direct action of antifascists as *"totalitarian"* and *"the real fascism."* The journal of the UK Revolutionary Communist Party, *Living Marxism*, would continue beating this anti-antifa drum for decades, until they reinvented themselves as the far-right *Spiked Magazine*, even publishing a defense of postmodernism as the necessary antidote to "wokeness." [145]

While it would break what's left of Jordan Peterson's rotting brain, there's a long and sharp history of antipathy between antifascists and authoritarian communists. The very strategies of decentralized resistance and bottom-up sanction used by antifascists and feminists that Peterson is so horrified by and derides as "authoritarian" has long been strenuously denounced by actual authoritarians and postmodernists.

The two pillars of modern antifascism are **No Platform** and **Three Way Fight**. The first, a refusal to socially legitimize outright lies or facilitate the propagandizing of those operating in bad faith and recruiting through shows of power. And the second, a refusal to collapse fascists into merely a tool of the ruling establishment or harmless fellow dissidents, but recognizes them

---

[145] Patrick West, "In defence of Postmodernism," *Spiked*, April 15, 2023, https://www.spiked-online.com/2023/04/15/in-defence-of-postmodernism/.

as a distinct pressing threat. While a minority of non-anarchist leftists have been involved in antifascism, these two core and defining planks are obviously tied to anarchist ways of thinking. Just as we know that a handful of insurgent anarchists can (and frequently do) change the world without seizing institutional power, so too can a handful of insurgent reactionaries. Because this fact is unavoidable to anyone involved in struggle against fascists, even the non-anarchist antifascists are forced to recognize that the fight is not against one totalizing megamachine, but *many* quite different enemies simultaneously, in a far more complicated network topology.

The modern terms of antifascism were formalized and popularized in part in response to 9/11 as much of the left continued the legacy of Foucault's embrace of the Iranian Mullahs by embracing islamists like the Taliban and ISIS as supposed allies against US imperialism,[146][147] but, in rough form, the insights they represent have repeatedly emerged in every antifascist group from the Maquis and the Battle of Cable Street on.[148]

Yet to many postmodernists who, like Peterson, subscribe to the warped "totalitarianism" analysis, such a militant rejection of any pluralism that "teaches the controversy" of Holocaust (and Holodomor) denial can only be classified as totalitarian and fascist. If you refuse to tolerate neonazi terrorists who are outside and against the ruling order, you're seen as functionally *on the same side* as that ruling order.

While postmodernists sometimes talk in their papers like anarchists about networks and the microphysics of power, this instinctive campism—a legacy of their roots in the authoritarian communist ideological swamp—is constantly visible in how they argue.

At the same time that the Science Wars were raging, Bruno Latour was defending Jean-Marie Le Pen's racism, outraged at the violence of antifascists in the streets, crying about censorship and sneering that science could not *prove* racial claims to be false.[149]

Meanwhile, some of Meera Nanda's most cutting remarks during the science wars were simply tracing all the ways Vandana Shiva was in bed with

---

146 k-dog, "Fifth Column Fascism: fascism within the anti-war movement," *ARA Research Bulletin*, 2 (2001).
147 Michael Staudenmaier, "Anti-Semitism, Islamophobia, and the Three Way Fight," *Upping the Anti*. 5 (2009).
148 For a detailed look on the catastrophic failure of early anarchists in Italy, who dragged their feet on coming around to No Platform and Three Way Fight see my review of *The Individualist Anarchist Origins of Italian Fascism*, "From Stirner to Mussolini: Review: The Anarchist-Individualist Origins of Italian Fascism," *Center for a Stateless Society*, March 28, 2022, https://c4ss.org/content/56480.
149 Emmanuel Marin, *Summary of Articles from Le Monde*, February 24, 1997, https://www.jwalsh.net/projects/sokal/articles/emarin.html.

hindufascists of the Bharatiya Janata Party and Rashtriya Swayamsevak Sangh—headlining their conferences, endorsing their projects, allowing her words to be echoed by them without criticism. To Shiva this sort of antifascist critique by Nanda was unthinkable. The neoliberal ruling order was such a totalizing threat that such alliances with fellow dissidents were *obviously* necessary. And *whatever happened to civil pluralistic debate?*

Indeed, postmodernists far predate Jordan Peterson in whining about "cancel culture" and feminism run amok. Beyond the rabidly reactionary screeds of Lamborn Wilson and his friends (like the snitch Bob Black) against feminism, the academic milieu repeatedly banded together in the most vicious ways. When professor Dragan Kujundzic was sanctioned by his college for sexually harassing a young graduate student,[150] Derrida went to war for his tribe and power network, trying to blackmail the college with the threat of removing a donated archive and imperiously sneering without a shred of dignity:

> "How can she claim to have the right to initiate such a serious procedure and to put in motion such a weighty juridico-academic bureaucracy against a respectable and universally respected professor?"
>
> Jacques Derrida, letter to the chancellor of UC Irvine, 2004

Sure, there's the jaw-dropping pomposity of the old fraud—offended that the hierarchies of academic prestige he'd invested his life climbing and reinforcing could be trumped by a mere girl—but beyond that lurks the old "totalitarianism" and campist framework. Derrida treats it as a trivial background consensus that the power structures of patriarchy and academic prestige, the actions of an individual predator, are not even worth considering against the True Enemy of an institutional juridical system.

This sort of rot is *systematic* in postmodernist circles. As previously mentioned, when Avital Ronell was revealed to have sexually assaulted, harassed and stalked a student, Chris Krause of *Semiotext(e)* instinctively stood with the oppressor rather than the oppressed. But she was hardly alone. Judith Butler, Slavoj Žižek, Joan Wallach Scott, Gayatri Chakravorty Spivak, Jean-Luc Nancy, Lisa Duggan, Jonathan Culler, and many more openly fucking defended her.

Echoing countless scumfucks in punk or metal whining about how they totally "support" antifascism or feminism as struggles against a distant enemy, but not in any case where it involves conflict within their scene, Diane Davis

---

150 Roy Rivenburg, "Were sex and punishment behind feud for archives?," *Los Angeles Times*, February 25, 2007 https://www.latimes.com/archives/la-xpm-2007-feb-25-me-derrida25-story.html.

wrote to *The New York Times*, "*It's so disappointing when this incredible energy for justice is twisted and turned against itself.*"[151] And Žižek scoffed in Ronell's defense that she was a "*walking provocation.*"[152] How could provocation ever be bad?!

Now let's be clear: rape culture and its patriarchal apologists are presently endemic to almost all ideological subcultures, certainly including many prominent examples in academic physics and analytic philosophy, and it is trivially the case that Judith Butler, for instance, supporting a predator does not somehow automatically invalidate other ideas they argue for. My point here is twofold: first, to peel back the pretension that postmodernist academic circles are supposedly more politically enlightened, but, second, to zero in on the way that the underlying "totalitarianism" thesis reproduces the most boring reactionary perspectives in the Left, and is leveraged to provide cover for enemies beyond The System.

Postmodernism doesn't reject binaries or universals consistently—that's trivially impossible—instead its "rejection" of such creates new sweeping binary metanarratives that reinforce the Left's preexisting tendency to view the world in terms of a single unified homogenous institutional enemy, and the necessity of pluralistically unifying all dissidents against it. Throwing Iranian women under the bus to support the fundamentalist revolution on the grounds of promoting *difference* against global uniformity is not out of place in the Left, rather it perfectly exemplifies how the majority of the Left has long been at odds with anarchism's universalist radicalism. A divergence that postmodernism accelerated.

Feyerabend is particularly illustrative of the reactionary currents popular in the Left, because of the sheer gulf between his appropriation of "anarchist" as a label and what he actually found worthy of praise in the Left. He showered approval on monsters like Marx, Lenin, Trotsky, and Mao, but would only sneer at the sole anarchist he referenced, Kropotkin.

> "Public action was used against science by the Communists in China in the fifties, and it was again used, under very different circumstances, by opponents of evolution in California in the seventies. Let us follow their example."
>
> Paul Feyerabend, *Against Method*, 1975

---

151 Alex Bollinger, "Why is Judith Butler trotting out tired excuses to defend a sexual harasser?," *LGBT Nation*, August 14, 2018, https://www.lgbtqnation.com/2018/08/judith-butler-trotting-tired-excuses-defend-sexual-harasser/.

152 Katherine Mangan, "New Disclosures About an NYU Professor Reignite a War Over Gender and Harassment," *The Chronicle of Higher Education*, August 15, 2018, https://www.chronicle.com/article/new-disclosures-about-an-nyu-professor-reignite-a-war-over-gender-and-harassment/.

I want to be clear that most of the doctors repressed by Mao's policies were closer to craftsmen or engineers applying practical skills than radical scientists probing for deep universal dynamics, and there was both tacit and experimental knowledge of value in traditional Chinese medicine, with the spread of things like ephedrine a massive boon to humanity. But let those caveats not eclipse that this talk of "public action" was Feyerabend endorsing Mao brutally rounding up doctors and sending them to the countryside to die while promoting folk medicine that Mao knew was not as effective and would not use on himself. Many of these doctors only survived because local peasants trusted their results more than the folk practices (a hodgepodge mixture inclusive of demonology and astrology) propped up by the regime.

Feyerabend's enthusiastic prescription for scientists is brutal centralized political violence, a frank statism he repeatedly happily endorsed, albeit sometimes under more indirect euphemisms like *"democratic control," "political means,"* and *"non-scientific controls on science."*

Feyerabend's approval for the truncheons of communism did not stem from any genuine progressive concerns. What Feyerabend endorsed was the centralized *top-down* state violence to defend any reactionary parochialism under pressure from rationality or evidence. And he had a sharp distaste for *bottom-up* establishment and enforcement of progressive norms as exercised by anti-authoritarians like feminists and anarchists:

> "There is much reason to suspect some of the elements of political correctness."
> Paul Feyerabend, *Farewell to Reason* (2nd edition preface), 1988

This is in the context of pushes in the '80s by activists to make verboten casual racism and misogyny, one of the most quintessential examples of decentralized anti-authoritarian methods working outside the centralized and authoritarian structures of democracy, where reactionaries were simply out-organized by boycott networks of altruistic activists willing to self-sacrifice harder to shift social norms. Feyerabend's framework inevitably means siding with the fascists whining about how our suppression or marginalization of them is "totalitarian." This reactionary instinct to prop up the obviously oppressive or false in the name of "diversity" is consistently visible in Feyerabend, with him even grouching about the Second Vatican's progressive reforms in the first edition of *Against Method*:

> "Whoever does not like present-day Catholicism should leave it and become a Protestant, or an Atheist instead of ruining it by such inane changes as mass in the vernacular."
> Paul Feyerabend, *Against Method*, 1975

Should it be any wonder the nazi pope, Joseph Ratzinger, was a fan and formally cited Feyerabend's work as *justifying* the Catholic Church's repression of Galileo? How much should we read into Feyerabend's catholic background?

What we see again and again is a stunning disinterest in critiquing real *physical* violence and massively centralized power, from Maoist China to the Catholic Church, much less any interest in active insurgent resistance to suppress racism and patriarchy. The proper struggle is instead, we are informed, one of metaphysical aesthetics and temperament.

If the earnest realism of scientists has the brash and crude stench of the uppity commoner compared to the more refined aristocratic pluralism of academia, that's intolerable *"totalitarianism."* And for such rudeness they should get rounded up and slaughtered by an all-powerful state like Mao's.

Multiple times during my undergrad, other activists who had learned I was majoring in physics expressed shocked revulsion and explicitly informed me I should be *murdered,* or at least forcibly reeducated for such, come The Revolution. A promise they sometimes delivered without a second thought. *"Oh, and have you read Lacan yet?"* Invariably they were middle class or some other subcategory of unbearably rich, and, just as invariably, their supposed political commitments evaporated upon graduation.

Of course, Feyerabend usually didn't let his allegiances shine through so directly. His writing, like so many postmodernists of the era, is drenched in references to distant oppressed peoples that he is nobly fighting on behalf of. Just as he was fighting for the faith healers refusing to take kids like me to the emergency room, he was fighting for some fantastical accounts of voodoo he read or the caricatures of indigeneity being promoted by Levi-Strauss.

Such was the zeitgeist of his era in academia, with noble savage narratives being all the rage in anthropology since the Man The Hunter conference in 1966. This lazy overcorrection for anthropology's genocidal white supremacist past, led to things like anthropologist Frédérique Apffel-Marglin expressing outrage at the introduction of vaccines to India, on the grounds that this was a violation of *local particularity,* waxing poetic about how traditional faith healing rejects a distinction between matter and spirit. At the apex of all this white psychodrama, Marshall Sahlins would infamously lecture down to Gananath Obeyesekere, defending the gobsmackingly racist claim that the Hawaiians really did see Captain Cook as a god. As all cultures must be totally incommensurate in their worldviews and thus to ascribe rationality or concern with objective truth to them is to become *The Real Imperialist.*

Crusaders in this zeitgeist weren't content with simply reinforcing dominant colonial narratives with the valences reversed, they had to even dig up and try to rehabilitate fringe reactionaries, with postcolonial theorists like Ashis Nandy and Shriv Visvanathan explicitly defending the occult medical practices of the theosophists Helena Blavatasky and Annie Besant (familiar to any antifascist researcher as influential precursors to the nazis) as "*ethnoscience at its most autonomous.*"[153]

Such orientalism was a rot in left academia that spread far broader than the postmodernists, but it was deeply entangled with postmodernism's entire orientation and metanarrative, to say nothing of their push against science. Lamborn Wilson's racist fantasies and fetishizations as "Hakim Bey" are infamous, but his years in Iran were spent as an editor for The Imperial Iranian Academy of Philosophy, an institution founded by Seyyed Hossein Nasr and the orientalist Henri Corbin on mysticism and Aryan racial supremacy that was all *directly lifted* from the Nazis. Lamborn Wilson's enthusiastic embrace of the term "postmodernism" a year later in New York cannot be separated from this.

> "The enemies of Sufism are… the intangible though very real forces of secularization and modernization which threaten tradition, the Tradition, everywhere in the world."
>
> Peter Lamborn Wilson and Nasrollah Pourjavady, *Kings of Love*, 1978

We can laugh at the racism of white kids dressing themselves up as defenders of "the third world" in deeply patronizing ways, like *Tel Quel*'s maoist cult, but it's worth emphasizing how such defense of the gods of "particularity" and "diversity" against universalism directly pulls from and empowers oppression and reactionary currents.

To keep up the anti-universalism metanarrative requires aggressively erasing and silencing all the examples of radical scientific practice, materialist worldviews, and concern for Truth that have arisen from below and are still struggling to overthrow their particularist oppressors.

Lamborn Wilson only traveled to Iran after being expelled from India for undisclosed reasons, and he was only interested in the history of these brown people to pillage mystical baubles, so who could expect the pederast trustfunder to bother learning anything about either society? But if he had given a fuck about the brown people being tortured and slaughtered around him, he might've appreciated the depth of local histories of atheist materialists, like the famous scientist Omar Khayyam. He might have learned that atheist

---

153 Nandy, Ashis and Shriv Visvanathan, "Modern Medicine and its Non-Modern Critics: A Study in Discourse," in *Dominating Knowledge: Development, Culture, and Resistance*, ed. Frédérique Apffel Marglin and Stephen A. Marglin (1990), 161.

materialists had been struggling against the priests and ruling castes in India for millenia, with even the Upanishads referencing them. The major school of Charvaka has long endorsed materialism, empiricism, and induction and derived its name from being "worldly" and "prevalent among the people."

Today, these ancient Indian materialist/atheist philosophers have proved central influences on the upsurgence of anarchism within Burma, with anarchists like Thiha JP centering their anarchism on atheist physicalism. In India proper, a vast number of activist groups form The People's Science Movement which have pushed realist positions against faith healing and champion activist causes around things like conception and AIDS, they were also the ones to expose the coverup of Union Carbide's mass murder in Bhopal.[154]

Who benefits when indigenous atheists in Canada, for example, find themselves aggressively defined out of existence?[155] Who benefits when the only resistance that can be acknowledged is mystical and spiritual? Who benefits when truth is discarded for *pragmatics* defined in terms of the "health" of some Community?

Again and again, over the decades, I have been told that one or another disruptive survivor in the activist Left should shut up about the fact of her rape, that her dogged allegiance to the truth at any personal cost is perverse and an act of treason to the stability of "The Community," without which we would have no hope of resisting greater evils.

Postmodernists have loved to abstractly reference "the subaltern"—those excluded from discourse, whose voices are not heard or even recognized. But when pressed in things like the Science Wars their instinct has consistently been to enthusiastically further such exclusions via the rhetoric of campism and orientalism. *Don't you see there's a greater enemy? We have to close ranks. It's only pragmatic.*

Who does this benefit?

> "Woman's skepticism, Nietzsche suggests, comes from her disregard for truth. Truth does not concern her... If Nietzsche and Derrida can occupy and speak from the position of woman, it is because that position is vacant and, what is more, cannot be claimed by women."
>
> Teresa de Lauretis, *The Violence of Rhetoric*, 1985

---

154 Roli Varma, "People's Science Movements and Science Wars?," *Economic and Political Weekly* 36, no. 52 (Dec. 29, 2001 - Jan. 4, 2002): 4796-4802.

155 Jonathan Simmons, "Indigenous atheism and the spiritual stereotype," *OnlySky Media*, June 29, 2022, https://web.archive.org/web/20220708172643/https://onlysky.media/jsimmons/indigenous-atheism-and-the-spiritual-stereotype/.

10

POSTMODERNISM'S INFLUENCES ON REACTIONARIES

"This is written into the Judeo-Christian doctrine, right from line one: It's through the discourse that we engender the world. And that is a divine principle."
>
> Jordan Peterson, *Sam Harris vs Jordan Peterson*, 2018

"We are now at the end of the Age Of Reason... There is no truth, either in the moral or scientific sense... Science is a social phenomenon and like other social phenomenon is limited by the benefit or injury it confers upon the community."
>
> Adolph Hitler (as alleged in *Hitler Speaks*),
> Hermann Rauschning, 1940

"From the identitarian's perspective, postmodernism's anti-foundationalist broadsides constitute an emphatic justification of tradition's particularity and the fact that we are who we are only because we make certain decisions to identify with and defend our particular system of truth. The constructed (that is, the human or cultural) character of the historical narrative, the multiplicity of these narratives, and their absence of closure are cause for affirmation and commitment, not despair, for culturally relative 'truths' born of one's own identity are necessarily more meaningful than those that are not."
>
> Michael O'Meara, *New Culture, New Right:*
> *Anti-Liberalism in Postmodern Europe*, 2013

It should be the least surprising thing in the world that reactionaries have found succor and inspiration in postmodernism.

Unfortunately many have little experience in wider political circles, encountering conservative positions only in mainstream media or faculty funding fights. If all you see of right-wingers is Jordan Peterson, Fox News talking heads, and debate-me-bro college republicans, it makes sense to assume that conservatives are opposed to postmodernism, almost by definition. Examples of conservatives or fascists praising postmodernism and drawing from it are thus written off as contrived and isolated exceptions.

But the truth is that postmodernism has been systematically and vastly embraced by both traditional conservatives and outright fascists.

In the classics departments and seminaries that crank out national columnists, lawyers, judges, and the christian intelligentsia, there was certainly no friction.

Literature departments certainly had no issue conjoining postmodernist fads with conservative politics. At Yale, the "gang of four" literature professors who pushed Derrida and deconstruction into Anglo-American literary criticism in the '70s were staunch conservatives. In the case of Paul de Man, when it was exposed that he had written a pile of pro-nazi and anti-Jewish propaganda during the Second World War, a number of prominent figures in the tradition of deconstruction tried to defend him against this cancellation, with Derrida whining at length. (In another hilarious example of his broad ignorance, Jordan Peterson cites these early postmodernists at Yale as the origin of "identity politics" in the US Left.[155])

The prominent conservative legal scholar and US federal judge Richard A. Posner crowed about the reactionary and conservative currents in postmodernism, while emphasizing skepticism and moral relativism as the conservative position.[156] Probably the most prominent conservative philosopher in the latter half of the 20th century, Michael Oakeshott, is frequently described as a "postmodernist" in his skeptical and nihilist moves, including by Rorty who

---

155 "Jordan Peterson Exposes the Postmodern Agenda (Part 1 of 7)," *Epoch Times*, uploaded July 4, 2017, YouTube, 1:00, https://www.youtube.com/watch?v=P-kNzYttjSHE.

156 Richard A. Posner, *Law, Pragmatism and Democracy* (2003).

heaped praise on him.[157]

Christian seminaries embraced teaching postmodernist philosophers, emphasizing the parallels with their own critiques of reason and "objectifying" science, see for example, *The Postmodern God: A Theological Reader*. While cable news conservatives may denounce postmodernism, christian intellectuals had no such qualms; before he was appointed pope, the nazi Joseph Ratzinger directly cited Feyerabend's texts as a justification for the church's suppression of Galileo.[158]

Jean-Luc Marion's awards-winning *God Without Being* is just one example of theologians embracing Heidegger and Derrida as early as the mid '80s. The explicit influence and praise of postmodernism is particularly strong in the Radical Orthodoxy movement, but there's been plenty of love in intellectual conservative christian spaces more generally. James K.A. Smith's enthusiastic defense of postmodernism as a justification for religion, *Who's Afraid of Postmodernism?: Taking Derrida, Lyotard, and Foucault to Church*, was a hit, garnering praise and awards from major evangelical mags like *Christianity Today*. And there's countless other examples: *Overcoming Onto-Theology: Toward a Postmodern Christian Faith. The Next Reformation: Why Evangelicals Must Embrace Postmodernity. What Would Jesus Deconstruct?: The Good News of Postmodernism for the Church*. Etc. etc. etc. Some authors mixed mild social progressivism in with the inherently reactionary framework of religion, but appreciation for postmodernist attacks on science and reason circulated far within christianity. As did its defense of particularity of tradition, one which allowed for framing christianity as an enemy of universalism, even to the point of advocating alliance with islamic reactionaries against the supposed totalitarianism of universalism.

Still others have seen it as a direct question of struggle against material reality in terms that starkly parallel Christian Science's line on matter as a seductive belief.

> "What do you mean when you say reality? When you say reality what are you talking about? And is it possible that reality is something we conjure here as vessels and conduits of the Divine if we have the capacity to somehow in the moment through practice disavow the strong gravitational (literally) pull of the material?"
> Russell Brand, on *The Jordan Peterson Podcast*, 2024 [159]

157 Richard Rorty, *Philosophy and the Mirror of Nature* (1979), 389.
158 John L. Allen Jr, "Ratzinger's 1990 remarks on Galileo," *National Catholic Reporter*, January 14, 2008 https://www.ncronline.org/news/ratzingers-1990-remarks-galileo.
159 "The Collective Unconscious, Christ, and the Covenant Russell Brand | EP 444," *Jordan B Peterson*, uploaded April 30, 2024, YouTube, 30:50, https://youtu.be/Rl5Z54aYA-E.

More broadly, the return to pre-modern virtue ethics in the '80s heralded by Alisdair MacIntyre has been vastly popular and influential in the conservative intelligentsia. MacIntyre fought with Rorty and, through that lens, denounced postmodernism as just another variant of modernity unwilling to fully shed materialism, but at the same time he had a huge impact in getting contemporary conservatives to embrace relativism and it's not uncommon to see a blurring in conservative writing where virtue ethics is spoken *approvingly* of as postmodern, nor is it uncommon to see MacIntyre in the recommendations of postmodernists (e.g. positively reviewed on release in *Social Text*). See also anti-feminist reactionaries like Christopher Lasch who was an early fan of Foucault and was praised in turn by him.

Today the neoreactionary movement has prominently seized the vice-presidency of the United States, via Peter Thiel's appointee JD Vance. But when Curtis Yarvin founded the neoreactionary movement in the early 2010s, it existed primarily on the fringes of rationalist and transhumanist circles, and was quickly banned from the rationalist *Less Wrong* discussion boards. This formal expulsion helped foster a lasting hostility towards rationalism and even science among many neoreactionaries, with many becoming early adopters of a "post-rationalist" identity. This conjoined a rush to mine any thinkers critical of rationality, realism, or the enlightenment project with a "trad" effort to embrace christianity. They found the existing christian intelligentsia that had embraced postmodernism as a defense of relativism and traditional particularity. Even Yarvin himself now endorses a "postmodern" approach to mythmaking around race science and critiques of the supposed limits of reason.[160] Influences of this have persisted in positive references to postmodernism in residual outlets like *Jacobite Magazine*, *Palladium Magazine*, and *The American Sun*.

Beyond US christians and tech workers, postmodernism found significant support in European fascist movements, from the France-centered *Nouvelle Droite* ("New Right") to the Russia-centered "National Bolsheviks" (the two constitute the primary funding and support networks for other fascist movements worldwide). Alain de Benoist formed GRECE in France in 1968 and solidified it during the heyday of post-structuralism. The result was a fascist organization praised by members as *"pagan rather than Catholic, postmodern rather than anti-modern."* [161] This sort of explicit and direct language is not

---

160 Alexander Raynor, "Meeting with the Father of Neoreaction,", *European New Right Revue*, February 12, 2025 https://nouvelledroite.substack.com/p/meeting-with-the-father-of-neoreaction.

161 Michael O'Meara, *New Culture, New Right: Anti-Liberalism in Postmodern Europe* (2013), 31.

uncommon. *Counter-Currents* founder and *Occidental Quarterly* editor Greg Johnson praises postmodernist philosophers, repeatedly arguing that deconstruction and their critiques of science must be seized by the right, and that Heidegger lays *"the outlines of a post-totalitarian, postmodernist New Right"* [162] while other authors call for a *"nationalist postmodernism."* [163] Americans may be unaware of these names because they never got fawning writeups in *The New York Times*, but these are *the* leading figures of the modern fascist movement. Alexander Dugin, the Heidegger scholar, thelemic occultist, fascist opponent of "totalitarianism," and grand poohbah of "multipolarity" (and repackager of Ivan Ilyin, who asserted Russians were superior because they supposedly rejected reason and facticity) dances back and forth from bemoaning postmodernity as too soft on modernism to gushing over postmodernist authors and praising his own *"Eurasian postmodernism."*

> "Modernity was a mistake, a crime... Now we are witnessing the end of modernity."
>
> Alexander Dugin, on the podcast of fervent postmodernist Thaddeus Russell, 2022 [164]

Richard Spencer has an academic background in Adorno, has repeatedly praised Nietzsche and Heidegger (alongside Evola) as the central theorists of fascism, pushed antirealisms in his initial appeals to libertarians, and has many times happily admitted that the white race is an arbitrary social construct (in ways echoing Spivak's "strategic essentialism").

> "A fact only has meaning because of the perspective looking at it... There is no fact that's going to destroy identity or make it irrelevant... Facts are lame."
>
> Richard Spencer, Q&A at Auburn University, 2017 [165]

The culture editor of altright.com, Jason Reza Jorjani, got a PhD in philosophy from Stony Brook focused on continental philosophers; he blathers on

162 Graham Macklin, "Greg Johnson and Counter-Currents," in *Radical Thinkers of the New Right: Behind the New Threat to Liberal Democracy*, ed. Mark Sedgwick (2019), 205.
163 Eumaios, "Objective Fictions & Subjective Realities: The Need for a Nationalist Postmodernism," *Counter Currents*, May 5, 2020, https://counter-currents.com/2020/05/objective-fictions-subjective-realities-the-need-for-a-nationalist-postmodernism/.
164 "Unregistered 193: Alexander Dugin (VIDEO)," *Unregistered Podcast,* uploaded January 7, 2022, YouTube, 7:43, https://www.youtube.com/watch?v=X2RxutaSGQU.
165 "Richard Spencer's Full Q&A at Auburn University," *The Auburn Plainsman Online (ThePlainsman.com),* April 19, 2017, YouTube, 14:08, https://youtu.be/g1JJA6UiEio.

about a combination of Heidegger, pragmatism, and the occult, arguing that nature eludes our grasp and even attacking "technoscience" in a presentation to the National Policy Institute.[166] Darren Beattie, the Trump speechwriter who palled with Richard Spencer at the Mencken Club,[167] wrote a doctoral dissertation on Martin Heidegger, criticizing mathematical objectivity and framing the nazis as a lesser evil than modernity.[168] Julia Hahn, Special Assistant to Trump in his first administration, came from a background of psychoanalysis and critical theory, explicitly justifying power on the grounds of Foucault's claim that it is inescapable[169] and crowing about Slavoj Žižek's endorsement of Trump.[170]

Nor are these somehow examples of rare graduate students slipping through otherwise tame departments. The deputy speaker and "chief ideologue" of the fascist AfD party in Germany, Marc Jongen, got his PhD in Heidegger under the notorious post-humanist Peter Sloterdijk whose various sharply reactionary views (on reason, racial genetic pools, and feminism) once got even Habermas to break the toxic civility of academia and denounce him as a fascist.

In more cartoonish, populist, and conspiratorial circles, the wildly popular fascist youtuber Carl Benjamin (Sargon of Akkad) has long self-described as a *"postmodern traditionalist."*[171] While Blake Smith, who praises the fascist known as Bronze Age Pervert,[172] likewise praises the *"unwoke Foucault"*[173]

---

166 Ronald Beiner, "The Conservative Revolution of the Twenty-First Century: The Curious Case of Jason Jorjani," in *Contemporary Far-Right Thinkers and the Future of Liberal Democracy*, ed. A. James McAdams and Alejandro Castrillon (2022), 193.

167 "Trump aide 'fired over ties to white nationalist event'," *BBC*, August 21, 2018, https://www.bbc.com/news/world-us-canada-45249154.

168 Daren Jeffery Beattie, *Martin Heidegger's mathematical dialectic: Uncovering the structure of modernity*, (dissertation, Duke University, 2016).

169 "Consciousness and Society," spi conference i, August 19, 2013, *SPI – The Society for Psychoanalytic Inquiry*, uploaded November 25, 2013, YouTube, 48:23, https://youtu.be/39FMKMdoM18?.

170 Julia Hahn, "Slavoj Žižek Says He'd Vote Trump: Hillary 'Is the Real Danger'," *Breitbart*, November 4, 2016, https://www.breitbart.com/politics/2016/11/04/slavoj-Žižek-vote-trump-hillary-real-danger/.

171 He has had "postmodern traditionalist" in his Twitter bio since 2022. https://web.archive.org/web/20221130025349/https://twitter.com/Sargon_of_Akkad. See also Carl Benjamin (@sargon_of_akkad). "The problem is that reality as we experience it might indeed be radically subjective, and until you can demonstrate that it isn't, then we are in a bind, aren't we?" Twitter, December 7, 2024. https://archive.is/Q5buu.

172 Blake Smith, "Bronze Age Pervert's Dissertation on Leo Strauss," *Tablet*, February, 15, 2023, https://www.tabletmag.com/sections/arts-letters/articles/bronze-age-pervert-dissertation-leo-strauss.

173 Blake Smith, "The unwoke Foucault," *Washington Examiner*, March 5, 2021,

and has encouraged readings of Roland Barthes as a weapon against trans people.[174] Among the chuds who stormed the US capital on January 6th 2021, was professor of philosophy and widely praised hipster musician John Maus, whose doctoral dissertation was packed with incessant references to Derrida, Foucault, Agamben, Deleuze, etc.[175] Mike Cernovich, infamous for his huge role in spreading countless conspiracies like "Pizzagate," put his inspiration frankly, *"Look, I read postmodernist theory in college. If everything is a narrative, then we need alternatives to the dominant narrative."* [176]

And pretty much any conservative, of any stripe, interested in discrediting science on some issue has embraced postmodernist arguments and the intellectual legitimacy they bring.

Figures from Christopher Caldwell, of the notoriously far-right *New York Times*, to R. R. Reno, of the conservative christian magazine *First Things*, heaped praise on Agamben during COVID.[177][178] The old Foucauldian had denounced distancing norms and referred to the virus as a "hoax." [179] But the love affair goes back further.

Philip Johnson, the creator of the "intelligent design" repackaging of creationism explicitly appeals to and praises his influence from epistemic relativism, postmodernism, and the strong programme.[180] The notorious theocrat-funded Discovery Institute that is central to pushing "intelligent design" incessantly echoes postmodernist critiques and complaints of "totalitarian science." Sheila Jasanoff, a scholar in STS/SSK, warned as early as 1996 that as a leading scholar she was being approached repeatedly by corporate interests to undermine the credibility of basic science, and being quoted in court to

---

https://www.washingtonexaminer.com/magazine-life-arts/2227707/the-un-woke-foucault/

174 Blake Smith, "The Inner Life of Gender," *Tablet*, February 15, 2023, https://www.tabletmag.com/sections/arts-letters/articles/gender-neutral.

175 John Mas, *Communication and Control*, (Ph.D., University of Hawaii at Manoa. 2014).

176 Andrew Marantz, "Trolls for Trump," *The New Yorker*, October 24, 2016, https://www.newyorker.com/magazine/2016/10/31/trolls-for-trump.

177 Christopher Caldwell, "Meet the Philosopher Who Is Trying to Explain the Pandemic," *The New York Times*, 21 August, 2020, https://www.nytimes.com/2020/08/21/opinion/sunday/giorgio-agamben-philosophy-coronavirus.html.

178 R.R. Reno, "A Responsible Bishop," *First Things*, April 17, 2020, https://firstthings.com/a-responsible-bishop/.

179 Benjamin Bratton, "Agamben WTF, or How Philosophy Failed the Pandemic," *Verso Books*, July 28, 2021, https://www.versobooks.com/en-gb/blogs/news/5125-agamben-wtf-or-how-philosophy-failed-the-pandemic.

180 Robert T. Pennock, "The Postmodern Sin of Intelligent Design Creationism," *Science & Education* 19 (2010): 757-778.

this end.[181]

When it comes to climate denial, Sven Ove Hansson has detailed a pile of examples of academic postmodernists, in particular in the STS tradition, directly defending climate denial in the '90s well after the IPCC consensus, including sharply complaining about reductionism and realism.[182] To give just his first two examples: Between 1990 and 1992 Frederick H. Buttel and collaborators Peter J. Taylor, Ann P. Hawkins, and Alison G. Power wrote a series of academic papers where they explicitly leveraged a variety of relativist frameworks like Bloor's strong programme to critique the idea that global warming is a true model of physical reality. They pulled from a right-wing think tank, praised deniers, described climate scientists in harsh terms, denounced Margret fucking Thatcher for listening to them, and constructed a variety of explanations in terms of social power for the global warming consensus, studiously ignoring (because of symmetry) any consideration of the actual physical facts. Warnings that such global environmentalism was aiding the forces of globalization were included for good measure. In 1993 William Cronon denounced a fellow historian for implicitly ceding the existence of global warming, "*as if that constructed knowledge had some objectively persuasive authority*" and sneering that someone might be "*a closet realist after all*" and ended up citing denialist literature from the Cato Institute. Hansson's whole work is rich with detail on the crossover between antirealist respectable STS academics and right-wing climate denialism, as well as their resulting ties and collaboration with right-wing institutions.[183]

And while there are obviously many causes behind increasing skepticism of science among conservatives, polling shows that conservatives are significantly *more* likely[184] to be skeptical of global warming if they have a college education. Such conjunction of college education with rejection of scientific consensus is a recent development not present before the '80s. This tracks with the proliferation of postmodernist arguments and assumptions in academia having a broad influence in terms of providing conservatives with justifications and a sense of support when they want to reject something. It's also the *opposite* of what you would expect from a tale of climate skepticism arising from media manipulation of the uneducated.

---

181 Jasanoff, Sheila, "Beyond Epistemology: Relativism and Engagement in the Politics of Science," *Social Studies of Science* 26, no. 2 (1996): 393-418.
182 Sven Ove Hansson, "Social Constructionism and Climate Science Denial," *General Philosophy of Science* 10, no. 3 (2020): 1-27.
183 ibid.
184 Frank Newport and Andrew Dugan, "College-Educated Republicans Most Skeptical of Global Warming," *Gallup*, March 26, 2015, https://news.gallup.com/poll/182159/college-educated-republicans-skeptical-global-warming.aspx.

This direct influence on conservatives has been commented on from within the remains of the postmodernist and STS ranks. Indeed some, like Steve Fuller who played a prominent role in the Science Wars, have even testified on behalf of creationists and praised the increasing widespread utilization of STS arguments by creationists and conspiracists as *legitimization* of STS, condemning the abnegations and faintness of will about getting into bed with reactionaries:

> "STS can be fairly credited with having both routinized in its own research practice and set loose on the general public—if not outright invented—at least four common post-truth tropes… What is perhaps most puzzling from a strictly epistemological standpoint is that STS recoils from these tropes whenever such politically undesirable elements as climate change deniers or creationists appropriate them effectively for their own purposes… We should not be quick to fault undesirables for 'misusing' our insights, let alone apologize for, self-censor or otherwise restrict our own application of these insights."
> Steve Fuller, *Is STS All Talk and No Walk?*, 2017

To be fair, this broader repolarization of conservatism has been facilitated in no small part by the rampant spread of the totalitarianism thesis in the '80s, well beyond postmodernism proper, with diversity frequently framed by generic liberals as the supposed antithesis to fascism. Fascists and the wider racist reactionary hordes were not unaware of the currency this narrative had in wider society and they aggressively embraced it. But since many of these narratives had been appropriated from the original nazis themselves, they did not have to work hard. The exact same historical arguments against globalism and ethics and technoscience were smoothly replicated with "totalitarianism" the central charge against them. The hereditarian racists in the IQ debate defined themselves as the "pro-difference" camp. Neonazi street fighters branded themselves as mere proponents of "human biodiversity."

This is a personal anecdote, but an arresting one: long ago I dated an ostensible anarchist, a Women & Gender Studies major—at the time into postmodernism—who, to my horror, abruptly went on a tirade about how they were a skeptic about the Holocaust because they were a skeptic of history in general and it was "totalitarian" to expect them to even believe the world existed before they were born and blah blah blah. *"I'm not saying it didn't happen; I'm just saying it's authoritarian to assert that it did happen."* While there was certainly wider antisemitism lurking behind this, it was also clearly reinforced by an instinctive need to spite all social censure or pressure. While Lyotard in his post-*Libidinal Economies* era deserves praise for resisting the erosion of facticity in the Faurisson affair, there is no doubt in my mind that the American postmodernist ideology that set down roots in the '80s, framing all dissent from establishment narratives as virtuous and transgression as the highest imperative, has helped billow the sails of holocaust denial, a point that Deborah Lipstadt makes at some length in her book *Denying the Holocaust*.

Let me be absolutely clear, because so many who came of age in the '60s and '70s seem fundamentally incapable of grasping this point, *every* ideology and social milieu has instances of fascist and reactionary creep, flows of appropriation and influence, but this is not reason to throw up our hands and ignore all connections and influences. It can be quite illuminating to examine such to find ideological weak spots and dangers. Further we have a moral and political obligation to aggressively confront and attack such sociological connections, to cut off the flows and publicly warn about infections, never to obscure or deny them. Instead, the default has been complete denial, sneering, gaslighting, and jaw-dropping minimization about the influence of postmodernism on reactionaries (and vice versa), and this has only facilitated the wider problem.

Across the wider conservative spectrum denialisms of various forms (Holocaust, climate change, evolution, etc.) grew into great prominence alongside the rise of postmodernism in the '80s and aggressively leveraged much of its metanarrative. While conservatives in the '50s had been happy to see themselves as hegemonic, squashing the upstart dissidents and weirdos, they have since largely embraced the zeitgeist pushed by the postmodernists: they demand "ideological diversity," they frame themselves as skeptics and rebels fighting totalitarianism and universalism.

Many comrades laugh about the claims that "conservatism is the new punk," but there is a sense in which the old white reactionaries in wraparound sunglasses sadly *do* better represent much of the old punk scene than Crass or Chumbawamba ever did. Certainly their unhinged rants in their pickups about "feminism as fascism" or the unfairness of not being allowed to say the n-word are indistinguishable from those of Bob Black. Punk icon Jim Goad, whose racist and pro-rape zines were once distributed right alongside Lamborn Wilson's zines—likewise praised for their "transgressive" nature—is now a superstar among the alt-right. The totalitarianism thesis, the hostility to any truth or hegemonic norm, might have had appeal as a weapon of resistance in a context of a stultifying technocratic monoculture, but it was an irresponsibly deployed weapon whose remains continue to poison the earth long after its use.

11

FUNDAMENTALISM AMONG THE WARRIORS

"The heart of the matter, we believe, is that many philosophers and sociologists want 'truth' and 'reality' to be exact. To them 'truth' cannot have an error bar. Reality cannot have any residual uncertainty."
Kenneth G. Wilson and Constance K. Barsky,
Beyond Social Construction, 2001

"Lay audiences who are unfamiliar with these fields are susceptible to being swayed by dumbed-down soundbites like 'sex not gender' and 'woman means adult human female'—phrases that sound pithy and compelling precisely because they oversimplify reality."
Julia Serano, *What is a Woman*, 2023

"Feminist doctrine invents several new categories of thought-crime, the most heinous of which is to question or contest current feminist doctrine [sexual harassment]… It shrewdly guilt-trips those campus liberals who will agree to all sorts of nonsense rather than let themselves be labeled misogynist."
Norman Levitt, *Why Professors Believe Weird Things*, 1998

Fundamentalism Among the Warriors 251

When Trump ran for office in 2016, the fascist mobilization was beyond anything in living memory—including the bad old days of the '80s, when hundreds of neonazis like East Side White Pride openly ruled the streets of cities like Portland, jumping minorities for sport.[185] Not only were many of those same old boneheads, some of whom had served time for murder, back on the streets, but they were joined hand-in-hand by ravenous hordes of conservative know-nothings. At first, a so-called "Fascist-Ancap Alliance" brought Infowars fans and armed militia members to chant "build a wall" on the Portland State University campus at students of color. Then a wider number of "street preachers" and "white identitarians" soon coalesced in streets, chanting promises to murder immigrants and beat non-white infants to death almost every weekend, then jumping in their trucks and driving around looking for victims. By May of 2017, one of their number, Jeremy Christian—who had a couple months earlier punched a bus stop beside my face, attempting to make me flinch—had stabbed three people on our light rail train, killing two, after attacking two muslim girls and screaming about a local antifascist research group.

During this siege of our city, while far-right calls for murder and intense state violence echoed outside his windows, one professor at Portland State, Peter Boghossian, chose to spend his time plotting revenge against some more dire enemies.

Boghossian joined with James Lindsay and Helen Pluckrose to compose twenty vapid articles (on things like rape culture in dog parks and fat bodybuilding) and submitted them to an array of peer-reviewed journals in queer, trans, feminist, fat, sexuality, and race studies. Four out of the twenty were accepted and published.

In the conservative media ecosystem—in between calls for mass deportations and the arrest of all political dissidents—this was trumpeted as proof of the vacuity of "left wing identity politics" or "grievance studies." Very soon their hoaxes were branded "Sokal Squared."

At first glance, the hoaxes could not be more different, and many, including Sokal, noted this.

Sokal's Hoax took aim at metaphysical and scientific antirealism, and was

---

185 Elinor Langer, *A Hundred Little Hitlers: The Death of a Black Man, the Trial of a White Racist, and the Rise of the Neo-Nazi Movement in America* (2004).

an essay structured around exposing and critiquing Aronowitz's own writings. The hoaxes of Boghossian, Lindsay, and Pluckrose took aim at… *academics writing about the structural oppression of minorities?*

Sokal sought to prove that a prominent marxist zine had no ethical compulsion against platforming explicit declarations of a reactionary philosophy. Boghossian, Lindsay, and Pluckrose sought to prove that… *you can get a few mildly silly research topics past lazy peer review in some bottom-tier academic journals?*

The only way one could possibly equate the two situations is if you believed that Sokal's Hoax had anything to do with peer review or thought postulating the existence of things like white supremacy and cisheteropatriarchy were on par with postulating that the physical universe doesn't exist and our minds make reality.

In short, this new hoax "scandal" had almost nothing in common with the antirealist contentions at the core of the Science Wars. Attempts to label it "Sokal Squared" by conservative media personalities was surely just stolen valor and pandering to chortling troglodytes on Fox News who think "minorities whine too much."

Anyway, why did Meera Nanda sign the Rowling letter?

THE BIGGER ENEMY

When *Social Text* groused, in the aftermath of Sokal's hoax, that the entirety of left academia was under attack, they weren't *just* trying to bring in allies to cover their asses. Norman Levitt and Paul R. Gross' bestseller *Higher Superstition* may have brought to wider public attention the ongoing wars over realism and antirealism within academia and the radical left, but it *also* explicitly tied conservative critiques of the radical left to critiques of postmodernism. This courting of the far right was a huge part of its breakout success.

While Levit and Gross postured at various points as centrists, liberals, or even leftists, in no remote sense were they anything other than far-right reactionaries. They whined about gender abolitionism and racial justice, while posturing as "neutral" on basic issues of oppression, in ways that prefigured the language of today's demagogues masquerading as "classical liberals": *"We refuse on principle to take sides in the dispute over the literary canon, in the fights over affirmative action, in the question of whether it is well to have 'studies' departments for subpopulations with a history of victimhood."* [186]

"Victimhood." Not oppression, "victimhood." I mean, what can even be said to such reactionary drivel besides *eat a fucking bullet*.

The explosion of popularity that *Higher Superstition* saw is functionally inseparable from the simultaneous explosion of interest in Richard J. Herrnstein and Charles Murray's *The Bell Curve*. Public discourse in mainstream media consistently conjoined the two books, tying those scientists denying race pseudoscience to those postmodernists denying science and physical reality wholesale. But this move wasn't constrained to conservative op-eds. Among the physicists, Alan Cromer, for example, approvingly cited *The Bell Curve* in one of the many resulting books published in the academic half of the Science Wars.[187]

Outright hostility to feminism, multiculturalism, and decolonial struggles was an incessant infection among the critics of antirealism and relativism. In the same article where Fromm demands Aronowitz confront whether his antirealism substantiates that of Christian Science he *also* whines endlessly about feminist cancel culture.[188]

The leftist science warriors had every opportunity to denounce these would-be allies, but many chose silence or enthusiastic complicity. Just as

---

[186] Norman Levitt and Paul R. Gross, *Higher Superstition: The Academic Left and Its Quarrels with Science* (1994), 9.
[187] Alan Cromer, *Connected Knowledge: Science, Philosophy and Education* (1997).
[188] Harold Fromm, "My Science Wars." *The Hudson Review* 49, no. 4 (Winter, 1997): 599-609.

the postmodernists choose silence or enthusiastic complicity with the more batshit antirealists in their camp.

There was "a bigger enemy," after all.

By 2005, Sokal's primary collaborator, Jean Bricmont, had come to the conclusion that the struggle against Israel's genocidal apartheid regime and the suppression of Palestinian voices licensed him to collaborate with fascists and conspiracy theorists. So pressing was the greater enemy of Zionism.

Soon he was declaring the genocide in the Balkans to be an invention of NATO propaganda and Milosevic a smaller enemy than the US.

Amidst scoffingly dismissing the support that leftists once lent Pol Pot and Stalin, Bricmont explicitly renounced the Three Way Fight perspective of anti-fascism, in terms that characterized the thinking of many leftists at the time:

> "The neither-nor stance gives the impression that we are somehow situated above it all, outside of time and space, whereas we are living, working, and paying taxes in the aggressor countries or their allies (in contrast, the position 'neither Bush nor Saddam' made sense for Iraqis, since they were subjected to both regimes). An elementary moral reaction would be to oppose the aggressions for which our own governments are responsible, or else to approve them outright, before even discussing the responsibility of others."
>
> Jean Bricmont, *Humanitarian Imperialism*, 2006

When, in 2009, antifascists exposed Bricmont's praise for the Holocaust denier (and Rwandan genocide denier) Paul-Eric Blanrue, he doubled down, denouncing antifa as well as anti-racism, even getting Chomsky to sign a public letter on behalf of another holocaust denier. Further and further he spiraled, and soon he was praising holocaust deniers more broadly as *"atheists of the religion of the holocaust."* [189]

> "For me, the world is split in two: those who defend Chouard [a denier of gas chambers] in the face of the Parisian media inquisition and the others. Everything else takes second place."
>
> Jean Bricmont, Facebook, 2019 [190]

And Sokal?

Sokal, whose face years ago I printed on militant anti-postmodernist stickers and buttons for anarchist bookfairs? Sokal, the leftist radical who taught for Sandinistas and was one of the early prominent academic figures organizing against the apartheid regime of Israel? Sokal, the childhood hero of a kid trapped in a faith healer cult?

---

189 Jean Bricmont Personality Notice, *Conspiracy Watch*, Last Updated 2021, https://www.conspiracywatch.info/notice/jean-bricmont.

190 ibid.

Alan Sokal took speaking engagement after speaking engagement with the far-right, repeatedly sharing platforms and heaping praise on them. When James Lindsay released a ludicrous and transparently reactionary book titled *Cynical Theories: How Activist Scholarship Made Everything About Race, Gender, and Identity*, Sokal endorsed it, even writing a preface.[191]

Lindsay would quickly move on to spreading fear about *"white genocide,"* [192] describing queer folks as *"a hostile enemy,"* [193] and alleging that shadowy global elites were conspiring to reduce the world population.[194]

But most directly, Sokal has embraced rabid transphobia and gotten deeply into bed with the most noxious fascists. He defends *conversion therapy* for trans kids,[195] citing the discredited pseudoscience of Ray Blanchard, the notorious crusader against survivors of child sexual assault. The field of sexologist conversion therapy studies is notoriously unscientific, going back to literal defenses of "faith healing."[196]

---

191 Alan Sokal, Preface to the French translation of Helen Pluckrose and James Lindsay, *Cynical Theories*, 2021, https://physics.nyu.edu/faculty/sokal/preface_to_cynical_theories_ENGLISH.pdf.

192 James Lindsay (@ConceptualJames), June 10, 2021, https://web.archive.org/web/20210628000113/https://twitter.com/ConceptualJames/status/1402749362220974094.

193 James Lindsay (@ConceptualJames), June 17, 2020, Twitter, https://archive.is/8geaj.

194 James Lindsay (@ConceptualJames), January 13, 2021, https://web.archive.org/web/20210324145446/https://twitter.com/ConceptualJames/status/1349370680857518082

195 Alan Sokal,. Submission to the Government Consultation on "Banning Conversion Therapy," December 5, 2021, https://physics.nyu.edu/faculty/sokal/conversion_therapy_submission.pdf.

196 David H. Barlow Ph.D., Gene G. Abel M.D. & Edward B. Blanchard, "Gender identity change in a transsexual: An exorcism," *Archives of Sexual Behavior* 6 (1977): 387–395.

An Interlude On Fascist Creep

As we've covered, there was a pile of reactionary shit among the ranks of postmodernists, many of their perspectives or arguments had directly reactionary roots, and much of their legacy was empowering conservative science-deniers and outright fascists. ...So the science warriors had garbage among their ranks too. Everyone's hands are dirty. What do we do with this?

Again, the easy path out—and one that many unfortunately instinctively swerve towards—is to declare all such ties as irrelevant. *Everything can be appropriated by reactionaries, anyone can be corrupted—so examples of such neither matter nor illuminate.*

This deflection is gravely mistaken because it abandons our obligation to identify *specific* points of commonality, crossover, and ideological weakness. It treats fascism as some inexorable storm that can swallow anything; a brute sociological phenomenon rather than an ideological cluster with content and structure.

Of course, such a dismissive approach is comforting to many and is quite common in academia, where surveilling rank-and-file fascists and the ideological moves that make up their various movements is at best seen as janitorial work for *mere activists* like antifa. At best academics engage with a few select historical fascist *academics*, while treating the arguments or connections of actually influential living fascist thinkers and organizers as almost inherently irrelevant.

But the compromise of *"I don't press about your camp's associations if you don't press about my camp's associations"* only ends up providing cover for fascists in practice. Similarly, *"Why should we focus on a few kooks rather than intellectually stronger or more popular positions?"* is democratic-brained naivety regarding the damage that a well-positioned minority can inflict or the corruptive influence it can have.

These dismissals spawn severe ignorance and misunderstandings, often outright endorsing the enemy's perspectives or setting up strawmen (like the "totalitarianism" thesis), and it can lead to practical miscalculations and failed predictions in the struggle to vanquish the fash, to say nothing of embarrassing pratfalls in direct debate or engagement with their ideas.

The fallback in the face of such is to scoff and shrug, *who among any of us could possibly predict or engage with this lunacy!* A solidarity in mutual ignorance of fascist thought is thus cultivated as a signature of elite belonging. The more that the in-group is—in their ignorance—*flabbergasted* by fascists, the more being ignorant of and flabbergasted by fascism becomes a virtue, critical to belonging to the in-group.

In reality while fascists come in many stripes, and reaction takes many forms, it does not take *every* possible form. No matter how diverse and seemingly preposterously cobbled together different ideological currents in the fascist movement are, there are commonalities and important underlying structures to examine and understand. So too with reactionaries and conservatives more generally.

Pressing such points is not about invalidating by association—or playing whataboutism—there is little shame in unchosen and happenstance proximity. The real demerits arise from inaction. As modern antifascists painstakingly learned during the '80s when neonazis first surged in subcultural scenes, we have a responsibility not just to draw lines and exclude reactionaries from the various prestiges and legitimizations accorded by association, not just to publicly expose all fascist creep and bust up their organizing, nor just to meet their jackboots in streets with defensive violence, but to studiously take an accounting of how their creep into various circles happens.

Every ideological community has vulnerabilities that can be exploited; how aggressive you are in discovering and applying patches over those vulnerabilities is what counts.

But if irrationalism, parochialism, and relativism are the traditional hallmarks of reaction—not universalism or belief in a single physical reality—what could even be the source of reactionary rot among the realists?

You Don't Have To Be A Scientist...

On October 12th, 2020, one Toby Young, the "general secretary" of a UK nonprofit he formed by the name of "The Free Speech Union" sent me a fancy letter implying they would sue a US-based anarchist organization I was a part of for "libel" over a public denunciation we had made of a misogynist and racist shitstain named Toby Fitzsimmons. The aforementioned shitstain had submitted a generic article on anti-capitalism to us earlier that year, only to be exposed by his fellow students at Dunham University, who admirably leaked a pile of screenshots from a misogynistic and racist chat he'd been in and gave us a heads up. Stating our opinion on the matter publicly—I was informed by this "Free Speech Union"—was a *"social media pile-on"* and thus constituted a *"criminal offense."*

I laughed and told the censorious little tory pricks to fuck off.

A few years later, Sokal proudly gave a talk before them.

For the most part, this talk was a lazy-retread of his decades-prior fame, recycling the same quotes and the same canned arguments he'd repeated in talks again and again. But Sokal added something new this time:

> "Saying that quarks are constructed is wrong but at least it's subtle, saying that sex is socially constructed is—I mean, you know, you don't have to be a scientist to realize that that's wrong."
>
> Alan Sokal, talk before the Free Speech Union, 2024 [197]

You don't have to be a scientist?

In his Hoax, Sokal groused (in mocking praise) that Aronowitz had written *"neither logic nor mathematics escapes the 'contamination' of the social."* Meta-commentary on this line continued for years throughout the Science Wars, with exactly how much contamination the central issue. I always read this focus on things like math and physics as a mutual recognition of what had the purest claim to science, the most objective structures of reality. It surely went without question that *everything else* was strongly contaminated with the social; fields like biology were repeatedly brought up as spaces of fuzzy distinctions and perspectival choices. Objective reality would have to live or die on truly *universal* structures, as painstakingly uncovered by the most radical sciences, not the half-baked impressions and pragmatic needs closer to everyday life.

Three decades later, before the censorious reactionaries of the "Free Speech Union," this would all be thrown out. *"You don't have to be a scientist."*

[197] "What is Science and Why Does it Matter? With Professor Alan Sokal," talk before the Free Speech Union. March 27, 2024, *The Free Speech Union*, uploaded April, 2024, YouTube, 55:54. https://www.youtube.com/watch?v=UbRrP8UzvSc.

The little tories didn't want to hear Sokal repeat the same canned snark and damning quotes he'd been regurgitating since the '90s; they only knew who he was because they'd heard he'd once owned the libs. So he tailored his talk into a denunciation of transfeminism. He knew what lines would serve as fresh meat to the jackals, and he tossed them out with relish.

> "Emmy Noether, one of the greatest mathematicians of the 20th century was denied a professorship at Göttingen because why? —Because she had two X chromosomes."
>
> Alan Sokal, talk before the Free Speech Union, 2024[198]

Let's pause.

It is *transparently false* that Noether was denied professorship in 1915 because of her chromosomes. Theophilus Painter wouldn't discover the X/Y heteromorphic chromosomal pair in humans until 1924; whatever Noether's chromosomes were, the philosophical faculty at Göttingen simply could not and did not discriminate against her because of them. They discriminated against her because she was a woman.

How, you might ask, did they determine this?

Well they no doubt assessed her physical characteristics as broadly aligning with a distribution of characteristics they associated with women, her build, her size, her fat distribution, her voice, the texture of her skin, and her lack of facial hair. They assessed how she dressed, how she carried herself, how she talked, and how she behaved. Most of all, they assessed how other people assessed her and how she assessed herself.

When Noether was later *expelled* from Göttingen by Hitler's ban on Jews, did the Third Reich assess her ethnicity by means of a DNA test and some arbitrary cluster grouping analysis of genes?

No, of course not. She was assessed as a "Jew" much as she was assessed as a "woman." In the rather loose associative terms of a concept constructed by humans through millennia of arbitrary social division, oppression, identity-formation, and malleable perpetuation.

Of course, it is broadly likely that Noether had two X chromosomes, just as it is broadly likely that she would have genetic markers common among Jews. But we do not *know* that, and certainly neither did anyone at Göttingen.

What *would* have been known to them was the existence of trans people. Karl M. Baer had undergone gender reassignment surgery in 1906 and gained *full legal recognition* in Germany with a new birth certificate as a man (a legal status that would have applied at Göttingen). In 1907, upon the death of prominent Brazilian socialite Dina Alma de Paradeda in Breslau, her geni-

---

198 ibid. 1:09:26.

tals were publicly revealed and a biography published. From these cases and others, German media had widely covered the existence of trans folks, and a surprising amount of the coverage was sympathetic. In academic spheres, Magnus Hirschfeld, whose prominent Institute for Sexual Science in Berlin would later be targeted and destroyed by the Nazis, championed the notion of *sexual intermediacy*. This notion, that sex was a multidimensional spectrum, followed in the footsteps of many scientists, like Charles Darwin, who had assumed the same.

And what we know *now* is that there are many people who develop the physical features we associate with women *without* two X chromosomes. Some have a Y chromosome but with genetic coding such that testes and testosterone never occur and they are born with the phenotypic traits we associate with women, like a uterus, fallopian tubes, cervix, and vagina. This often happens via a variation within a gene (called "SRY") on the Y chromosome *or* via a gene (called "SOX9") on a completely different chromosome, not traditionally considered a sex chromosome. Other folks with Y chromosomes have genetic coding on their X chromosomes such that their androgen receptors studiously ignore that hormone, causing them to develop the phenotypic traits we associate with women. Other genetic variations that lead to a female phenotype include having a single X chromosome, having a small piece of the Y chromosome in an X chromosome, or having both XY and XX cells from the merging of twins in utero.

Most women—with or without these traits—do not know anything about their genetic code. And certainly no one alive in 1915 knew these details.

Sokal dismissed out of hand such facts about the microphysics of "sex," in terms that are *astonishing*:

> "The sexual divide is an exceedingly clear binary. As binary as any distinction you can find in biology... a subject that's been more or less accurately understood by humans albeit without all the scientific details ever since the beginning of our species."
>
> Alan Sokal, speech before the Free Speech Union, 2024 [199]

---

[199] ibid. 24:01.

RADICAL ACCOUNTS OF "SEX"

> "If the ordinary things of the world are compounds then it is natural that they should share nothing in common except resemblances... Scientifically significant generality does not lie on the face of the world, but in the hidden essences of things... it is clear that 'table' and 'red' are not real universals; and 'gene' and 'molecule' are."
>
> Roy Bhaskar, *A Realist Theory of Science*, 2008

Some will say that the efforts of people like Sokal to redefine our everyday lay concept of "women" in terms of chromosomes is "reductionism." Perhaps in one sense of the term. It certainly strips away a vast amount of dynamics and declares them either derivative of a single underlying fact or irrelevant. But *lossy* reductions, ones that deliberately ignore further underlying dynamics and paper over the resulting exceptions, are myriad and infinite. They are not in line with the "reductionism" seen as a virtue and pursued by physicists.

Why would we ever expect the crude conceptual structures and terminology of early humans (assuming they even used a sex binary) to be preserved in a more radical and universal account of reality? And why would a physicist treat concepts at the highly coarse-grained level of chromosomes as *more real*, and somehow *less abstract* or *less prone to artifacts of social construction* than at the level of genes or, in his shocking example, quarks?

After three decades as a prominent champion of realism, it turns out Sokal is not fighting for the radically realist perspective justified by theoretical physics but rather an fundamentalist "realism" closer to something like a phenomenological or "common sense" perspective, where whatever *seems* to be the case *must be*.

For comparison, let's return to Maxwell's electromagnetism and Weinberg's electroweak theory. Physicists rightfully boast that the theory of electrodynamics has been proven precise to 10 in a billion. It matches all the data we have to the maximum precision we can test. Still, it is probably the case that this present theory is a rough or incomplete abstraction over a deeper reality and a more universal theory. We do not yet know this more radical account, and there's a tiny chance that it might not involve objects/relationships like "the electron field" as we know it. If so, we will be forced to do one of three things, depending on what the deeper theory looks like:

1. We could dissolve the concept of "the electron field" and never again use it for anything serious or even in any context. As folks did with "phlogiston."

2. We could continue to use "the electron field" as a shorthand, but recognize that it is useful only in some context or bounded domain, and tied to the pragmatic interests, perspectives, and applications of humans. It might retain some faint realist echo, as representing some degree of fixedness or factual structure, but it would not be *universally* valid.

3. We could try to redefine "the electron field," mapping it onto our new conceptual structures, *choosing* some specific mapping so that the term retained its universality.

Once upon a time we had a rough concept of "sex," one that emerged from pragmatic concerns with the technology of reproduction. It was a rough and crude bundle of associations, shaped in a feedback loop with social and cultural dynamics and pressures, but it seemed to point to some deep underlying reality. Something fixed in the world, a universality discernable from any perspective or vantagepoint. A single bit that God flips when creating a soul. As it turns out, this is absolutely and thoroughly not the case.

One can sympathize with the binary hypothesis. For a moment it seemed like a subset of the things we had historically clustered under the concept of "sex" were the result of a single chromosomal bit, XX or XY, and it is reasonable that some folks felt an urge to redefine "sex" to mean exclusively those things, potentially splitting off everything else once associated as mere cultural constructions of "gender." This gender stuff could then be messy and arbitrary, while sex is left clean and binary.

But as time went on the picture became complicated, biologists trying to preserve the binary hypothesis were forced to split their concept of "sex" into layers: *fetal chromosomal sex, fetal gonadal sex, fetal hormonal sex, internal reproductive sex,* and *genital sex* roughly corresponding to an idealized temporal sequence of development. A given individual could then be classified as male in one layer and female in another.

For any *scientific* concept of "sex" it's natural to want the broadest possible compression, a concept that could be generic across species. It would be weird to limit sex to "humans"—an already notoriously arbitrary category—much less assign particular genes like SRY and SOX9 a universal role.

But chromosomes and reproduction don't work the same way across nature. Birds encode sexual expression via a vastly different chromosomal mechanism from humans, requiring *double Z* chromosomes to encode the development of what we then call male features, and there are plenty of species that don't have a *chromosomal* sex binary at all. There are lizards and insects that don't use males at all and insects with three sexes. There are 7 taxonomic families and 27 orders of fish species that change sex, with female fish transforming into fully reproductive male roles. The causality of this process goes from social cues to the brain to the hypothalamus and pituitary gland which then cause the gonads to change. These fish and many other species reproduce through causal mechanisms that oblige blurring or rearranging the layers of analysis we might want to break them into for epistemic ease. Social and environmental context is often influential, as are an organisms' cognitive response, after all evolution will select for genes that respond to relevant external conditions and this involves being able to change their expression.

Every scientist can recognize that such *ad hoc* distinctions are the sign of a degenerating research program, a radical account that once seemed within grasp, slipping away. Most scientists would give up on preserving the more abstract term of "sex," either retiring it to casual shorthand and different definitions in different contexts, as biologists have, or dissolving use of it entirely.

In contrast to electrons, "sex" has always been a descriptor of the loosest common associations smeared out over the most absurdly differentiated set of examples; an obviously pragmatic and perspectively-oriented conceptual tool. It would be preposterous and *alarming* to any realist worth their salt if such a human-scale concept actually did cut reality at the joints because such would suggest that the structures of reality had been somehow tailored to our perspective.

If physicists had found a deeper zoo of particles beneath electrons (the chromosomes to sex), and then a zoo of particles beneath those (the genes beneath the chromosomes), we might've tried to hang onto "electrons" and speak of something like different isotopes of electrons dependent upon different underlying configurations.

But the problem is far more intractable in the case of "sex." The number of genes is *many orders of magnitude* larger than particles in the standard model or atoms in the periodic table of elements. And they're not universally fixed; the set of genes to be considered varies across individual "organisms," and even within them (imagine electrons needing to be defined differently on different planets). Moreover the arrangement of these genes matters, and the fuzzy notion we're trying to isolate, "sex," is a *process* involving massive interactions, with relevant causal chains often seemingly skipping up levels of abstraction in terms of reach (imagine electrons responding to the fact that it's Tuesday or that the cute lab tech frowned at you this morning). In short, "sex" is *nothing like* the universal perfect consistency of the electron field, where every single electron is indistinguishable from every other electron, modulo external arrangements like position and spin entanglement.

If physicists discovered that what we had been describing with "electrons" were somehow the product of such a batshit Rube Goldberg contraption and so massively and arbitrarily variable we would instantly give up on the concept of electrons. We wouldn't even throw them in the phlogiston bin, we'd drive them to the city dump under the cover of the night and bury them in hopes no one would ever find our shame.

And on our way back we'd probably have a psychotic break, because the sheer extent of the consistency and interlocking confirmation of everything to do with electrons *up until that point* would oblige us to believe in something like a Deceiver of vastly superior resources and/or the pointless irrelevance of all thought.

Biologists themselves recognize that "chromosomal sex" is an aggregate abstraction over far more complex genetic dynamics, and many now follow this to the inevitable conclusion: that the notion of "sex" motivating this whole affair, *cannot be reductively mapped*, as it is a human-scale abstraction created for pragmatic reasons and thus inextricably entwined with culture.

> "Plentiful data and analyses support the assertions that sex is very complex in humans and that binary and simplistic explanations for human sex biology are either wholly incorrect or substantially incomplete. For humans, sex is dynamic, biological, cultural and enmeshed in feedback cycles with our environments, ecologies and multiple physiological and social processes."
>
> Agustín Fuentes, *Here's Why Human Sex is not Binary*, 2023

Pretty much no trans or intersexed person on the planet[200] is denying that what we call *Homo sapiens* has so far had a roughly bimodal distribution of sexual characteristics largely causally carried by XX and XY chromosomes and associated hormonal regimes, that reproduction has operated through differentiated gametes with sperm cells fertilizing ovum. Let me say that again, the actual physical *reality* is simply not being contested by transfeminists in these discourses.

Just as recategorizing Pluto as a "planetoid" rather than a "planet" (and then back again) wasn't an attack on reality and didn't undermine anyone's knowledge of the distribution of matter in the solar system, this is a changing of *labels* on our map of the structures of reality, not a denial of them. What's being fought over is the use of a *word*. And only antirealists in the vein of the wackiest postmodernists and new agers think that words constitute reality.

The most obvious and natural reductionist approach is to throw out the term "sex" for anything firmly universal, while retaining variations on it in specific perspectival and pragmatic contexts, continually emphasizing the

---

200 The only possible exception I know of is some fringe speculative writing by Lu Ciccia, in *Contra el sexo como categoría biológica*, which saw harsh criticism and pushback from Mexican trans folks. I have not yet acquired a copy.

contingency of these uses. In general the reductionist, or radical, prefers to speak in terms of more fine-graining realities like *"people with uteruses," "people who primarily run estrogen,"* and even *"people with double X chromosomes."* In medical studies such finer conceptual categorization allows for more explicit clarity around potential causal dynamics and entailments.

Of course this change of language changes our attention, it emphasizes the objective lack of solidity of "sex," how the aggregates we're vaguely gesturing at are bimodal not binary, contextual not universal. But these are also not up for debate scientifically; they are objective realities.

Insofar as there could be any rancorous disagreement, it has to be over the virtue of reductionism, that is to say radicalism itself.

Fundamentalism: Realism Devoid of Radicalism

Immediately after Sokal's hoax, Barbara Epstein appropriated the public attention and umbrage to declare that Judith Butler's *Gender Trouble* was a noteworthy example of postmodernist *antirealism*.

> "Butler argues that the conventional view of sex as consisting of two given, biologically determined categories, male and female, is ideological, and defines radical politics as consisting of parodic performances that might undermine what she calls 'naturalized categories of identity.' Her assertion that sexual difference is socially constructed strains belief. It is true that there are some people whose biological sex is ambiguous, but this is not the case for the vast majority of people."
>
> Barbara Epstein, *Postmodernism and the Left*, 1997

Epstein did this right after complaining that postmodernism threatened to undermine marxist class analysis. Now, she was certainly right to denounce the tendency among postmodernists to evacuate class from their social theorizing on the grounds that *"all interpretations or constructions of class interest are equally possible and equally valid."* I've repeatedly emphasized that postmodernism's pluralism is a classed phenomenon, as is its allergy to the existence of simple facts or truths that might oblige action rather than unending open interpretation.

But I think the conjunction of the complaint about undermining class with the complaint about undermining sex is revealing.

It is trivially the case that any taxonomy of class will be arbitrary and lossy. If we let them, leftists will spend centuries furiously and fruitlessly debating the boundaries between "aristocracy," "bourgeoisie," "proletariat," and "lumpen." Are homeless queer teens shoplifting, scamming, selling drugs, and trading sex "proles" or "lumpen"? Are cops "workers," albeit "traitors," or are they in some entirely different category? I don't fucking care. Why would different splittings of such crude taxonomic terms matter to a crew of said homeless teens on their way to burn down a precinct? Our evaluation of the underlying power dynamics would be the same.

How we cut up such aggregate abstractions only matter if you're assigning them an ontological *primacy*.

When Epstein scoffed that the *"vast majority"* of people are not intersexed, the anarchist and the physicist are left completely perplexed. The anarchist feels no allegiance to majoritarian democracy; we would give our lives to defend a minority of one. And the physicist knows that even a single exception destroys a universal. Radicals simply cannot parse *"vast majority"* as an argument with any compulsion.

But for the *fundamentalist* this is powerful rhetoric.

The right-wing libertarian Michael Huemer is a prominent proponent of "phenomenal conservatism"—the claim that we are epistemically justified in believing whatever *seems* to us to be the case, in the absence of evidence against it. While Huemer recognizes the occasional need to inquire and reason to strange conclusions, he champions a foundationalism to justification that inevitably bleeds into a lack of drive to hunt for deeper realities under appearances.[201] Huemer has a tendency to embarrass himself by believing whatever incorrect and vapid accounts of antifa or *"woke social justice warriors"* pervade his right-wing social circles and media diet, feeling no compulsion to do a deep dive looking into what we actually argue and do. He's the sort to write, *"According to most progressives, the main source of racism is Republicans and white people, not Democrats or black people,"*[202] as a supposed gotcha where many Democrats and black "leaders" supporting drug prohibition somehow disproves its racism (when in fact everyone on about the racism of the drug war constantly screams about them).

If, on first glance, something feels wrong or stupid, he regularly decides it probably is and delves no deeper.

In this, the star philosopher of phenomenal conservatism faithfully represents legions of 13-year-old white boys captured by reactionary influencers since Gamergate by pandering to their assumptions and instincts.

If it *feels* racist that antiracists treat white nationalism as distinct from black community organizing or personal biases, then it is racist. Why should you feel an obligation to spend years reading up on the history and continued influence of white supremacy? Why should you recognize the value of a definition of "racism" not as just personal bias, but in terms of broad and often diffuse power asymmetries, incentives, and alliances? And if you're Huemer, why shouldn't you cite that an entire 14% of US respondents (one in seven!) still openly oppose interracial marriage as somehow proof that racism is basically dead?

If it *feels* like a surrender of epistemic diligence when feminists use the shorthand "believe survivors" as a slogan then it must be. Why should you

---

201 By contrast to such, the position that *there is no foundation*, rather we are all thrown into the mix and eventually forced towards universal epistemic attractors can be seen as a kind of infinitism, but it really is a position that our starting priors or starting notion of epistemic virtues *do not ultimately matter*. It's an image where we're thrown out to sea at various locations and *have* to swim aggressively to eventually find purchase towards the infinite horizon. Huemer quite evidently feels no such frantic desperation, and this enables him to treat exploration as a more of a token virtue, to be dabbled with as one feels.

202 Michael Huemer, *Progressive Myths* (2024)., 94

think more richly around how to correct subtly inoculated incorrect priors, consider which personal heuristics can correct currently broken tendencies in mainstream social epistemology, or take personal responsibility for judgement and action beyond the crude measures of a courtroom? Why should you investigate rape and sexual assault statistics broadly when you can note a single math mistake in a single paper and then try to slice away the rest of its numbers by simply scoffing at modern definitions of rape and sexual assault?[203]

If it *feels* like the police state is broadly just on matters of race, then it must be. And if you make a long chain of inferences around evidence in cases of murder that take for granted at various points assumptions like liberal fantasies about how the police operate and that no one with conceptual schemas, thought processes, or allegiances influenced by a broadly white supremacist society *could have voted for Obama*... well the end result sure feels like confirmation.

Both Huemer and Epstein are content with their surface impressions of the world in the absence of overwhelming and immediately available evidence to the contrary. This is what Epstein meant when she brought up the *non sequitur* that intersexed folks are a *minority* of the population.

Trans and intersex activists regularly bring up that almost all atoms in the universe are either hydrogen or helium, with *everything else* making up only the tiniest sliver by comparison. *Atoms are binary*, they laugh, *any exceptions can surely be ignored or discarded*. But this mockery stems from an instinctively physicist way of viewing the world. Only radicals feel the nagging urge to dig into the exceptions, to push beyond some effective domain of "good enough" to find deeper root dynamics.

In the face of the existence of a mere minority, Epstein felt no such compulsion. The fundamentalist mind feels the reverse: any existing models must be *defended*. This inevitably extends to violence.

There's a quote of Sokal's from right after the Hoax that has always eaten at me.

> "There is no fundamental 'metaphysical' difference between the epistemology of science and the epistemology of everyday life. Historians, detectives and plumbers—indeed, all human beings—use the same basic methods of induction, deduction, and assessment of evidence as do physicists or biochemists."
> Sokal, *What the Social Text Affair Does and Does Not Prove*, 1997

---

[203] No joke, Huemer expresses annoyance that things like *"repeatedly pestering someone for sexual favor"* can create sexual assault, and dismisses accounts technically describing rape but not using the term on the grounds that *"the simplest explanation would seem to be that respondents said they were not raped because they were not raped, in the ordinary English sense."*

In one sense, I agree wholeheartedly. I think it's obvious that no line can be drawn between the induction used in formal modern institutional science and the induction used by infants and everyday folks. This is why I argue that skeptical attacks on "science" can't help but extend to object permanence; there is simply no sustainable distinction between scientific antirealism and metaphysical antirealism.

But we apply these tools in different contexts, on different subject matters. Reductionism works wonders in physics, yet flounders in less radical fields studying messy aggregate systems where our concepts don't cut at objective joints but rather *contextually useful* distinctions. Sokal's quote in 1997 ignores this, which has long troubled me. By his talk in 2023 he has *reversed* things. Biology becomes *more obviously real* than physics, "sex" more real than quarks, because it is closer to the common sense of 13-year-old white boys.

If an account of the world makes a broad compression, who cares if it's a lossy one?

When I say that everyone engages in scientific thinking, I mean to emphasize the ever-present potential for radicalism. *"You too can plumb deeper! Many terrifying structures larger than us have no independent power or autonomous existence! They are ultimately illusions that can be shattered! We only have to press beyond their effective domain!"*

But it turns out what Sokal was trying to do was borrow the prestige of science for everyday beliefs. It wouldn't be the first time a socialist dabbled with reactionary populism, telling their caricature of an Average Joe that he already knows enough and *"to hell with anyone trying to fuck with that."*

Reactionaries always believe that some comforting simplicity must be defended from any remainder it leaves out.

Sometimes this comes guised as an epistemic conviction, like in Scott Adam's conspiratorial assertion that, *"Complexity is always a cover for fraud. In every domain."*[204] But simplicity-for-simplicity's sake is the goal. Not finding strange new lossless compressions of reality, but violently avoiding the demands of thought. While a scientist happily admits when something in our daily lives is actually irreducibly complex, the genesis of reactionary thought is the refusal to tolerate such. As a neoreactionary once memorably put this overall philosophy in a discussion on how to more fully oppress women: *"An exception must die so that a rule may live."*[205]

---

204 Scott Adams, Twitter, February 7, 2025, https://archive.is/ivPoh.
205 Pseudo-chrysostom, comment on "The Disastrous Effects Of Females In Power," *Jim's Blog,* June 5, 2018, https://blog.reaction.la/economics/the-disastrous-effects-of-females-in-power/#comment-1782721.

> "The denial of complexity is the essence of tyranny."
> Giorgio Parisi, *Nobel Lecture: Multiple equilibria*, 2023

In his attempt to defend the crude middle school account of sex against a more radical or reductionist account, Sokal is quickly obliged to abandon any universal definition, speaking instead of "humans" and mammals. One could imagine him defending this on the grounds that "*Homo sapiens*" are of greater import, he's just trying to give an objective account of something in *human nature*. Ignore all the ideological alarm bells going off and don't listen to any of those dirty punk transhumanists emphasizing that the open mutability of "humans" to processes of thought and culture has long defined us better than any arbitrary set of genes.

Sokal clearly wants to make chromosomes the defining core of human "sex." So he throws genes like SRY and SOX9 under the bus of his account of the True Human, to say nothing of androgen insensitivity, etc. He repeatedly emphasizes that intersex deviations are "*syndromes*." By this means one can dismiss the existence of intersexed folks as examples of the chromosomal sex mechanism *breaking*.

The fact that a starkly feminine model and jazz singer like Eden Atwood could grow up virtually indistinguishable from other women and treated by patriarchy in exactly the same way before randomly learning she had a Y chromosome, is irrelevant; Sokal must force her into his category of male, just *injured, disordered*.

> "An infertile individual with a Y chromosome is still male, just as a one-legged person remains a full member of our bipedal species."
> Alan Sokal, speech before the Free Speech Union, 2024

But notice what Sokal is forced to accept. The very absurd *teleology* that Aronowitz ranted about in his wingnuttery is baked into any framework that speaks of "broken" or "failing to work." To think in such terms is to take an anti-reductionist top-down approach, whereby some outline at the macro-scale that we have imagined for subjective cognitive convenience becomes instead an *objectively intended* outcome, and a given fact at the micro is disrupting the plan, a gear in a machine that is failing to do its job in service of "the organism" or even "the species." Broadly speaking, this sort of top-down thinking is precisely what Aronowitz wanted to resuscitate. His crusade against the reductionism of physics was all about rehabilitating the top-down and "holistic" conceptual approaches he thought were essential to marxism and critical theory.

To make the assumption that some united organism—like the working class—*exists* as an ontological primitive rather than a subjective epistemic shorthand papering over the finer-grained account, is to engage in anti-reductionist claptrap on par with Dupré or the vitalists. The same sort of thinking, writ large, would lead one to think that *nations* exist beyond and above individuals, such that an individual who works against "the will of the nation" or "betrays" it, should be labeled *"dysfunctional."*

Why should we consider a gene that doesn't set off a causal chain to build a specific set of gametes or whatever as "failing" or being "broken" because there's some larger aggregate system—or account of such that we've invented—we might grant moral authority to?

Many genes in our bodies never trigger or are suppressed, indeed this is often vital to what we subjectively consider our flourishing. We like some genetic dynamics and dislike others, but for deeply perspectival reasons that would not necessarily apply from the point of view of an organ, cancer cluster, alien mind, or even just another human with a different goal.

Genes are closer to a vast ecological system than a sequence of discrete commands from God detailing some teleological blueprint, some divine and objectively written collective *purpose* of all our cells and atoms.

To write off the "exception" of intersexed folk, Sokal is forced into leveraging the same sort of irreduction, where it's not that "sex" is a crude "seeming" that is ultimately shredded by the knives of reductionism, but rather that "sex" represents an ontological force outside and beyond the actual biological facts, providing an *intended ordering* of the world.

The political valences of this fixation on preserving order and common sense against the "chaos" opened up by radicalism are obvious.

> "The rigid authoritarianism of conservatives does not simply enshrine certain social norms, but always seeks to naturalize them, to uphold them as truths that are beyond the reach of society's capricious influences. There is no better mechanism by which one can validate traditional norms than by holding them to be unimpeachable, natural in some divine, ontological way that precedes politics, that precedes society. Gender can be made-up bunk now, since all the half-crazed transsexuals go around claiming it is—but sex, don't you dare deny the reality of sex! ... The core of conservative existential terror is the idea that social norms, even deeply-held, highly-embedded ones, could be questioned, altered, transformed—transcended. Even if they are someone whom the rules don't currently benefit, what if the changed rules benefit them even less? A cruel, tyrannical god may harm you, abuse you, but at least if you understand his rules, you can play them, you can attempt to curry favor. What could be more frightening than a world with no gods and no masters?"
>
> Talia Bhatt, *"Sex is Real": The Core of Gender-Conservative Anxiety*, 2024

In the video of his talk, Sokal stands before his audience of reactionary dorks and titters about feeling more welcome with them than among the Left.

Sokal is not alone. As the influence of anarchism has risen in the Left, aggressively changing norms and social expectations, a number of old school socialists have defected to the far-right, along with a number of postmodernists. But both are dramatically eclipsed in numbers by a set of former feminists, who abandoned pretty much every value they might've once held in favor of ravenous hatred of trans people. Collaboration with christians busy outlawing abortion? Easy. Paying neonazis for security at their rallies? A win-win.

Unlike Sokal, many of these reactionaries realized very early on that the root facts of biology were not on their side. To preserve "chromosomal sex" as a universal reality they would have to appeal to some ordering magick above and beyond science.

Guess what former arch-antirealist Sonia Johnson is now a realist about?

> "Males, therefore, constitutionally unable to meet the two-X-physical / spiritual-chromosome requirement, are forever prevented from being female."
> Sonia Johnson, *The SisterWitch Conspiracy*, 2010

Ordering The Army

Everyone knows that focusing on chromosomes is beside the point. That's not what is at stake in these questions of realism, radical or reactionary.

In Sokal's account he only became invested in arguing about sex and gender after talking with some older women of his generation who thought their political gains were being eroded by the "gender ideology" of the Kids These Days. The individual he cites most approvingly on these supposed political stakes is Kathleen Stock, a middling academic who had focused on subjects like aesthetics, film, and music before finding a more prominent career as a reactionary crybully and crusader against free speech.[206]

Stock's trade is now to assure those who do not get basic philosophical arguments around sex and gender, that she—a philosopher, by some technicality—likewise does not get them.[207] Infamously, she accused Judith Butler of necessarily having to reject the existence of things like *"molecules, atoms or quarks"* as an extension of their critique of *"sex."*[208]

In Stock's account of the epistemic virtues of science there are some notable skews: A good theory should be "simple" and "explain" the evidence, but only "well." She throws out anything like a virtue of lossless compression for instead the reactionary affinity for *lossy* compressions—simplicity at the cost of ignoring exceptions. As a result, her account is one stripped of any drive for deep roots. And you will not be surprised that she incessantly uses the word "radical" as a contemptuous slur.

> "Does a given theory explain the evidence well? Are there rival theories that might explain the evidence better? Does the theory help us explain and predict what people care about? Does it have other explanatory virtues such as simplicity, and is it a good fit with other existing productive theories?"
>
> Kathleen Stock, *Material Girls*, 2021

Of additional note is the frank subjectivism she places front and center: *"What people care about."* Despite her pretenses of reductionism or materialism and frequent declarations that she's defending "basic facts," Stock is flagrantly perspectival and pragmatist about how she thinks all our concepts should be

---

206 Stock infamously rage-quit her job over getting criticized for her lazy transphobic blather, and then got the British state to fine her former university over half a million pounds for having permitted students to protest her. The free speech of her critics was Orwellianly framed as somehow violating free speech by "bullying." She should be bullied forever.
207 Christa Peterson, "Kathleen Stock, OBE" *praile*, January 21, 2021, https://www.praile.com/post/kathleen-stock-obe
208 Kathleen Stock, *Material Girls: Why Reality Matters for Feminism* (2021), 63.

defined. She even approvingly cites the supreme anti-reductionist John Dupré on how labeling choices in biology reflect present human interests. Which, fair enough! Stock's core issue is that she believes these "interests" are being eroded by the alternative conceptual carvings of sex and gender long promoted by anarchists and feminists.

I strongly agree that the rough patterns we choose to pick out in aggregate macroscopic systems are unavoidably going to be shaped by our interests. We look at lumps of matter and identify "tables" and "table saws" not because those clusterings and distinctions are objective, but because carving up the underlying atoms that way is useful to us. Of course this doesn't mean that "tables" and "table saws" are scientific concepts or that such really exist! Like many physicists, I happily admit that the dynamics gestured at by the relativist sociologists apply to massively aggregate "objects" like societies and organisms; my argument for realism is that such get winnowed away once you get down to the universalities described by physics.

But Sokal wants to have it both ways. He wants to say his enemies have given up on objective reality, only to then embrace Stock's argument that our models *shouldn't* strive for objective reality but rather pragmatic and subjective utility. He wants to lean on the epistemic authority of physics, but then throw out its radical approaches to prop up crude approximations and teleology in biology. He wants to present science as a neutral arbiter of fact and those he disagrees with as *ideologues*, but then at the end of the day his argument is that we *must* hold onto a rickety concept of "sex" for political reasons.

It's impossible to divorce Sokal's reactionary privileging of "sex" over exceptions (and even quarks!) from his ideological commitment to a very specific analysis of patriarchy and supposed stakes to how we define "sex." This is how he can accept the adulation of a mob of misogynist tories and still imagine himself a feminist hero, a Hilbert standing for Noether against the patriarchy.

> "Hilbert fought strenuous battles with the university authorities to allow a woman to become a member of the faculty. A majority of faculty members argued against this: 'How can it be allowed that a woman should become a professor? ... What will our soldiers think when they return to the university and find that they are required to learn at the feet of a woman?'"
>
> Leon Lederman and Christopher T. Hill,
> Symmetry and the Beautiful Universe, 2004

But a radical realist would emphasize that the *exact* pathways of causation matter. The humanities professors—primarily philologists and historians—who blocked Noether's appointment instantly decided that the spouseless and childless prodigy with strong facial features was a woman because people *said* she was a woman, because in a thousand ways she affirmed rather than disputed this, and because she kinda looked like "women" were supposed to look.

Emmy Noether is an old hero of mine and it's *torturous* to have to break down the most simple facts of the oppression she faced in such basic terms. It's beyond enraging that Sokal would grab someone so dear to physicists and so casually strip out the *actual mechanisms* of her oppression. Noether's chromosomes—whatever they were—didn't create patriarchy. And patriarchy does not operate on karyotype tests.

Of course, reactionaries like Stock disagree.

In Stock's account, patriarchy is a product of biological sex: Noether's own chromosomes didn't create the social dynamics that oppressed her, but human sex chromosomes *in general* did.

> "Did a random group of people start oppressing random others? Or was it rather that there was one group better able, on average, to dominate the second group, due to genetics and associated tendencies to relatively superior physical strength? And if that, then how could the oppression or dominance itself have 'created' such characteristics?"
>
> Kathleen Stock, *Material Girls*, 2021

I want to be clear, I think pretty much everyone agrees it's probable that physical differences in bimodal sex expression—primarily issues of physical strength and pregnancy—helped break social symmetry. We see patriarchal societies over the vast majority of the planet, and it would be weird for this commonality to be the product of a single ideological *coin toss* over 60 thousand years ago, and an unbroken chain of ideological reproduction since. Whatever stochastic variation human societies have gone through, there is at least some statistical skew towards a preferred orientation of power across the bimodal sex expression.

What is not clear is how big of an impact this influence is. It may be very small or weak, merely capable of facilitating an initial symmetry breaking in certain egalitarian societies, before much stronger dynamics of technology, culture, ideology, identity, and institutions take over. Even the smallest perturbation can serve to catalyze the growth of massive crystals, yet not be particularly important to an account of said crystals. A small amount of dust may *set off* crystallization in a solution, but isn't that relevant in describing the actual chemicals catalyzing, the structures they form, or the forces at play between them.

Further, as anarchist historians repeatedly emphasize, we do not know how egalitarian societies (or matriarchal ones) survived and flourished in the holes in our records and between patriarchal empires. We know that many did, but there are systematic biases around what evidence would be left in the archeological record. How many other, stranger, systems came into being around something like sex/gender, whether oppressive or liberatory? How much did different expressions of technology, culture, ideology, etc intensify or oppose the influence of the bimodal trait distribution that Stock sees as central?

Note everything we have to gloss over in coarse-grained talk: The actual particular interactions of particular humans for millions of years. And in particular, the vast tangle of important mechanics in play today.

Radical feminists have long since settled on an account where patriarchy is primarily *upheld* by dynamics of ideology, culture, and institutions, even if biological differences in the bimodal sex distribution continue to play a part. Stock's argument for biological primacy is ostensibly reductionist, but it doesn't go through. She has to ignore the ongoing actual causal network and merely point to a roughly plausible temporal "first cause."

Stock wants a *simple* account of patriarchy, just as she repeatedly demands a *simple* account of gender or sex. But what we should care about in its dynamics cannot be mapped down to some objective reality of physics and remain simple. Feminists have long known that what we call "patriarchy" is closer to an *ecosystem*, a vast churning network of causal dynamics and congealing regularities. How could you define an entire *jungle* succinctly while providing a good map of it for our interests? Any parsimonious definition will bring some things into focus and occlude other things.

And just like the processes and layers of abstraction we might want do not cleanly separate with biological sex, gender does not cleanly separate from sex. It's all a mess. Gender is conditioned, inflicted, performed, experienced, habituated, incentivized, emulated, interpolated, invented, stumbled into, but so is sex, and the two are intermeshed. Patriarchy takes some *inspiration* and *influence* from the bimodal sex distribution, but is not reducible to it. It is not causally separate from dynamics of genes and hormones, but rather intersects with them in countless ways, a flowchart that cannot be summarized in any book.

This irreducibility reflects the fact that "patriarchy" is not a firm part of the ontology of the universe on par with electrons and quarks—it's perspectival *shorthand*, quite pragmatically useful in the subset of possible perspectives and situations in which we presently live. "Patriarchy" is in some sense real, but only like how "sex" is. Recognizing that is important because it also means that it is a concept that can fray and dissolve. Patriarchy is neither magically more than its components nor autonomous. It is not "natural" or an eternal fact; it has an *effective domain*. We can fucking kill it. We can make the world such that it fails to refer.

A huge part of this involves sabotaging its perpetuation of the sex/gender hierarchy.

> "Just as the end goal of socialist revolution was not only the elimination of the economic class privilege but of the economic class distinction itself, so the end goal of feminist revolution must be, unlike that of the first feminist movement, not just the elimination of male privilege but of the sex distinction itself."
> Shulamith Firestone, *The Dialectic of Sex*, 1970

> "Our ways of living and expressing ourselves break such fundamental rules that systems crash at our feet, close their doors to us, and attempt to wipe us out. And yet we exist, continuing to build and sustain new ways of looking at gender, bodies, family, desire, resistance, and happiness that nourish us and challenge expectations."
> Morgan Bassichis, Alexander Lee, Dean Spade,
> *Building An Abolitionist Trans and Queer Movement with Everything We've Got*, 2011

This isn't a blanket commitment to making ourselves "illegible" as a goal in itself, nor does it even remotely mean abandoning material struggles around from biology like abortion to matters of economics like unpaid housework. It doesn't mean *ignoring* distributions of matter in reality or taking choices of self-constitution or association away from anyone. It means actively looking for choices and moves that wouldn't appear in the conceptual frames native to patriarchy, but that are possible nonetheless. It means being insurgents or hackers, always on the lookout for vulnerabilities and opportunities, even when those are highly particularized. It means nurturing meshed prefigurative subcultures that do not wait for the slower "masses."

But to the reactionary mind, all this complexity is horrifying, unthinkable, intolerable. This isn't a campaign platform. This won't fit on a bumper sticker.

At one point Stock demands a *simple* definition of "man" or "woman" on the grounds that the simpler the definition the easier to translate between languages and cultures. But how would you even begin to describe what a "goth" or a "punk" is to someone outside your discursive community and frame of reference? How would you describe "race"—given that it doesn't fucking exist in any deep scientific or ontological sense—but also has immense practical existence as a feedbacking social reality. You would have to trail on for a long while, trying to account for every association, their weightings and contingencies.

What stands out dramatically in both Sokal and Stock—even though Stock references a number of actual feminists to mock them—is how they feel no urge to either investigate or think through the complexities of a different worldview. They expect everything handed to them in the lowest common denominator of public discourse. Stock's book is a veritable parade of hilariously surface-level readings and misinterpretations, even describing non-binary identities in terms of androgyny. It's a lot like reading Jordan Peterson or James Lindsay; there's a persistent reactionary entitlement to only the most immediate and simple explanations, protecting them from any worldview that would require too many points of change.

> "This approach finds its high point—or low point, depending on how you look at it—in academic Chloë Taylor's argument—in a feminist journal—that rape-crisis centres for women serve only to 'reinscribe gendered constructions of male sexuality as dangerous and of women's bodies as sexually vulnerable', and are 'the cause of rape', perpetuating the problem they seek to avoid. Frankly, this is mad."
>
> Kathleen Stock, *Material Girls*, 2021

When I was seven, my mother made a sharp choice in physical reality and abruptly packed her car, taking me and drove my sister to a domestic violence shelter. Upon arriving, she discovered they had a rule against housing boys aged eight or older; in a relatively short time I would be classified as an inherent sexual threat and my mother would be forced to choose to live on the streets with her children or to abandon me.

Such rules have long since been abandoned almost everywhere, with modern domestic violence workers speaking about this shameful past with horror. It's easy to understand why. If you force mothers to make a choice between sacrificing their children and sacrificing themselves, a great many will choose to suffer their abuser rather than give up their child to something worse. Often this means the continued abuse of mother *and* child. Over decades of such policies across the US, it's probable that hundreds of thousands of mothers were pressured into staying with their abusers. Many no doubt were murdered as a consequence.

Similarly, in the example of rape-crisis centers for women that Stock scoffs at, it should be *obvious* how such gender exclusion can increase rape, as cis women are not the only victims of rape, and denying support for cis men, trans men, trans women, non-binary folks, etc. increases the precarity and vulnerability of certain targets of rapists. A long time ago, when a male friend of mine was raped by an ostensibly feminist woman, she *bragged* about it in front of our mutual friends the next day, because she knew at the time there was no support apparatus, either social scripts or resources, he could turn to. Excluding a population from support eventually sends clear signals to predators. Every cop knows they can rape homeless trans kids without consequence.

My partner worked for years at a crisis center, answering harrowing calls and holding the hands of battered survivors in emergency rooms while the pigs glared at them. Decades prior, *in the bad old days*, that same crisis center had appointed as their director a woman who had horrifically abused her husband, and—after he finally escaped her—stalked his new partner, assaulted her, and then shot him in the genitals on the porch of his new house. At the time this was not considered abuse or disqualification from running a crisis center. Because she was a woman. In her new role she was—surprise—abusive to her staff, degrading the effectiveness of the entire crisis center, until she was finally forced out.

These are some of the most plain, object-level entailments of reinforcing gendered narratives where men are innately sexually dangerous and women are innately vulnerable, but there are less tangible if still substantive consequences. Hearing, as a small child, that I was innately dangerous, ontologically an abuser-in-waiting by virtue of my genetics, made the extreme abuse my mother continually inflicted on me and my sister hard to conceptualize and seemed to guarantee that no one would listen or recognize it. On other individuals such biological essentialism might gestate into fatalistic self-perception and self-narrative.

One can argue about the degree to which such dynamics play out, and I certainly disagree with much of Chloë Taylor's *broader* Foucauldian analysis, particularly around the way her account coddles rapists as "lacking community" and writes off vigilante action against them.[209] But, reactionary that she is, Stock never once thinks to trace any causality by which gender exclusions around support for survivors could hurt survivors. Simpler to declare Taylor "mad" and misrepresent her actual words:

> "In approving of rape crisis centers and ride services because 'they are run exclusively by women,' Woodhull reinscribes the construction of an adversarial and biologically grounded male/female dichotomy rather than challenging it. It is these constructions, I suggest, that are the cause of rape, and tactics such as Woodhull's may only perpetuate the problem they seek to solve."
> Chloë Taylor, *Foucault, Feminism, and Sex Crimes*, 2009

Stock ignores Taylor's more objectionable arguments to instead zero in on this because such constructions are *natural, obvious, common sense* to her.

When I once happened to relay the anecdote of my seven-year-old experience in our first shelter and its exclusion rule on Twitter—careful to note

---

209 William Gillis, "What's In A Slogan? 'Kill Your Local Rapist' and Militant Anarcha-feminism" *Human Iterations* (2024), https://humaniterations.net/2024/06/25/kylr/.

I'm really glad the shelter existed at all—hundreds of reactionary supposed "feminists" arrived out of the blue to deliver threats, demands that I kill myself, and general seething hate. It was only natural, obvious, common sense to them that eight-year-old boys *were* ontologically evil, disgusting, a waste of flesh, *"just as dangerous as grown men."* Most were outraged to learn that domestic violence shelters have long since renounced such policies. Others were outraged to learn "male" children of any age were ever allowed into such shelters. They thought it only natural that inclusion of a single small child would innately mean the exclusion of *"dozens of battered, bloodied and traumatized women and girls."*

When Stock positions genetics in such a core causal role to patriarchy, she is not merely talking about slight advantages in physical resources available to different sides of a bimodal distribution should things go sour. She is talking about a continuously propelling mechanism of such centrality and strength as to define the whole thing. A causation actively sustaining and reinscribing patriarchy at the level of values and motivations. One that would apply to a tiny child, raised entirely in a far-left feminist bubble, playing *Uno* on a shelter floor and drinking juice out of a plastic dinosaur cup.

The position of the self-described "feminists" screaming that such kids should be preemptively killed—that any woman would naturally and correctly see them as a threat—is that "males" rape and dominate because of their *genetics*.

This is a model of patriarchy not as an ecosystem to be disrupted and reconfigured, but as two warring armies.

One in which men and women are discrete classes, ontologically given facts of biology. And patriarchy is not something abstract and multifaceted in culture, ideology, and relationships as feminists try to encapsulate with "the rule of the father" or "the sex/gender hierarchy," but dirt simple: the supremacy of "biological men" over "biological women." There is no "third sex" built into patriarchy's function, where exceptions to the sex system are created and thrust beneath both men and women, to serve as whipping girl sacrifices to give men absolute free reign over and use as examples to keep women in line, to give them a stake in maintaining the system. No, no, no, there are just two armies.

For gender reactionaries sometimes too charitably called "TERFs,"[210] it's all utterly simple: one army is winning. For the other army to win it must

---

[210] It is unfortunate that "Trans Exclusive Radical Feminist" caught on as a liberal nicety, because it grants that these reactionary shitstains are feminists, much less radicals. Certainly the feminist movement has historically included some noxious currents including transphobes, but at a certain point "feminist" simply ceases to apply in any reasonable sense and quite obviously their thinking has nothing to do with "radicalism."

have *discipline*. Traitors (trans men) must be suppressed. Infiltrators (trans women) must be expelled. Party clarity and cohesion must be obtained. Power captured and leveraged.

> "Anti-trans feminists seek to still the category of women, lock it down, erect the gates, and patrol the borders."
> Judith Butler, *Who's Afraid of Gender*, 2024

> "We'll send Party militants to seize the key supply routes into patriarchy, then surround and storm the buildings that patriarchy is using as headquarters. After securing the surrender of the main body of patriarchy's army, we'll move out into the countryside to put down any provincial patriarchist counterrevolution. Is that the plan?"
> Ran, Facebook comment during the SF anarchist bookfair war, 2013

I want to be clear that I would never deny that there are enemies. Patriarchy is not a faceless external thing that just happens and we're all equal victims of. There are individuals deeply invested in it who have to be fought, resisted, and killed. And there are broad commonalities in society; the ecosystem of patriarchy constructs "men" to identify with it and provides benefits to them. Nor would I ever argue that there is no utility to collaborations and discussions closed to those of a similar experience.

But just as patriarchy is not centralized or monolithic, neither are effective resistance movements. We build a better world through distributed fluid insurgencies, where radicals don't wait for orders from above, but autonomously map the complex causal dynamics they are immersed in and leverage often quite particularized exploits. By understanding that our more universal enemy is the root dynamic of power, anarchists are always ready to recognize mutations or shifts in its operation, without having to tack on epicycles.

Authoritarians, whether leninists or liberals, miss this because of their ideological commitment to statist thinking.

I emphasize this political perspective because Stock's obvious reactionary commitments pervade every argument she makes. Her entire notion of resistance to patriarchy is in terms of "women's rights"—defined in terms of state legislation around exclusion and welfare. She violently simplifies arguments for trans people into discrete and completely separable planes of identity, social roles, and biology and (when all of them are obviously simultaneously at play and interwoven) and then judges her strawmen on their fringe exploitability as a *legal* framework. She rants about the watered-down over-simplified statements of academics and liberal NGOs, while studiously ignoring the moldy zines of radicals where all serious discourse and theory happens.

While there have always been some reactionary currents within feminist ranks, it's honestly hard to find anything recognizable as feminism in Stock's ideology; her worldview is so systematically statist and conservative. She whines that feminists are too pro-abortion and don't consider the wishes of the pregnant person's extended family.[211] She whines that a show shouldn't have been canceled because Greg Wallace is a sex pest and racist. Certainly she has nothing in common with the radicals that planted bombs in banks, burned down the houses of politicians, and nearly killed the king of Britain.

Leaders of the gender reactionary movement like Posie Parker (Kellie-Jay Nyishie Keen-Minshull) have already *explicitly* rejected identifying with "feminism" on the grounds that they think patriarchy is more or less exclusively a biological inequality rather than a self-replicating social and ideological system. Their prescription is social democracy: the existence of biological differences legitimizes the centralized violence of the liberal state to intervene in society to "protect women" as a disadvantaged group, restoring the balance. And as biology means that these interventions must continue forever, so too must the state.

Just as liberals consistently end up siding with fascists to crush anarchists and preserve the state, the gender reactionaries consistently side with the foulest of misogynists and fascists to preserve the patriarchal ontology of sex. *Both* have a vested interest in beating the belief into everyone that patriarchy is natural, so foundational is that premise to both of their projects.

From this perspective anything that might actually attack or undermine the state/patriarchy is a ghastly threat. It's akin to rejecting the toppling of the racist police state because without the police state how would we have affirmative action?

At one point in *Material Girls*,[212] Stock breathlessly compiles five headlines she's dug up since 2016 from far-right British tabloids involving lady criminals who are also trans. In her mind, a law against plastering headlines with someone's trans status is brainwashing society into misattributing male violence to women. Three of her examples are of child predators (apparently besmirching cis women in a way that hundreds of cis women child predators over that period didn't). But you'll never believe the other two examples she mixes in as comparably outrageous:

---

211 Kathleen Stock, "The perils of reproductive extremism," *UnHerd*, June 23, 2023, https://unherd.com/2023/06/the-perils-of-reproductive-extremism/.

212 For a sampling of other theorists tearing apart Stock's philosophical ignorance and sheer incompetence, see "While Tables Burn: On the (Non) Existence of Trans People and the Failure of Philosophy" by Talia Mae Bettcher and "The Matter Of 'Material Girls': Conspiracy Theory In Anti-Trans Philosophy" by Kim Hipwell.

> "'Gang of women repeatedly stamp on man's head in 2am brawl at Leicester Square underground station' (Daily Mirror website, 26 June 2018)… 'Woman who once shoved policeman onto Tube tracks jailed for spitting at officer' (Daily Mirror website, 17 February 2020)"

You read that correctly. Girl gangs fighting back aren't feminist. Spitting on or attempting to kill cops isn't feminist. Real women would never do such things.

This is not revolutionary feminism, this is just straight up reactionary authoritarianism. A craven worship of the state. In other words, liberalism.

12

THE TEPID LIBERALISM OF IT ALL

"A free society is a society in which all traditions have equal rights and access to the centers of power."
Paul Feyerabend, *Science in a Free Society*, 1978

"Words are never just 'words.' They are also speech acts. Thus, in addition to their semantic and denotative properties, words also entail perlocutionary effects, pragmatic consequences for human action."
Richard Wolin, *The Leprosy of the Soul in Our Time*, 2022

"If you're a real philosopher then the content of your work should be so deeply central to how you live that any criticism of it feels like a direct personal attack. If you're even capable of having a calm discussion, find another job"
Florence Bacus, Twitter, 2023

When one of the reactionaries of the Free Speech Union asks about abortion, Sokal prevaricates and stutters. Scientists should discover facts, he says. Questions of values should be decided by "democracy."

The same ideological concept that *Social Text* appealed to.

This is worth emphasizing. While both the science warriors and the postmodernists repeatedly tried to wrap themselves in pretenses of antiauthoritarianism or even full anarchism, almost everything about them was a product of enthrallment to this authoritarian framework: *democracy*.

So many postmodernists in the Science Wars wanted to place questions of science and truth up to a vote, sometimes literally to the point of asserting that physical reality itself directly sprang into existence via our collective beliefs. But the science warriors were not free from the political frameworks and class habits of liberalism either. In particular, they constantly made an implicit equivalence between *realism* (in terms of persistent and accessible non-mental structure) and *discursive reason* (as a socially communicative practice).

Liberals see within "realism" the promise of social convergence through democratic persuasion *without conflict or (non-state) violence*. Since this is their ultimate goal, they can't help but prematurely leapfrog towards it and equate being a realist with being a pacifist who solves every problem by means of civil debate, with deference to institutional meta-rules.

Sokal started off guns blazing on very stridently political grounds, but rapidly conceded too much: he explicitly said he was motivated by the catastrophe of postmodernism as a movement and its colonization of the Left, but he backed away from actually engaging with it *as a movement*. He singled out a small list of academics and then confined himself to merely pointing out they misunderstood or misused science. And while the postmodernists wailed that he wasn't civil enough, from the perspective of a radical all his choices were studiously civil to a fault, privileging the norms and self-image of academia. He never utilized the tools needed to disrupt, counter-organize, and bust up a *movement*, much less engaged in direct confrontation or action. He merely treated postmodernism as a series of bad ideas to be debated or mocked. Worse still, he tried to build a broad coalition like an electoral campaign, in many ways appealing to the lowest common denominator. He could have prosecuted his war in the journals, publishing houses, zines, and infoshops of the radical left, but he largely stuck to national newspapers and increasingly reframed his words in appeals to moderation and the center (and, eventually, the far-right). In short, he acted like a centrist or populist pursuing votes in the broad political body, not a political radical.

What is distinct and notable about the so-called Science Wars is how totally unlike a "war" they were. The science warriors never remotely acted as if the stakes were actually real. *Surely* if we were whisked back to the '90s—knowing the dire consequences to come in creationism, climate denial, and general fascist empowerment—extreme action to stop the postmodernists would be *morally obligatory*. And yet at the time not even a single package bomb was mailed.

I've appealed to examples of violence throughout this book to draw out these dynamics, to make explicit the political choices and assumptions the academics on both sides were making, and mock how transparently ridiculous their appropriations of "antiauthoritarianism" and "anarchism" were. This is not to *fetishize* violence as some universally superior tactic—in writing a book I demonstrably think discourse and persuasion are quite important and often preferable—but to highlight the irrational aversion to taking ideas seriously enough to act, a toxic norm in the fringe ideological subculture known as academia.

Liberals support intense violence—indeed they support the greatest engine of violence ever invented, the state—but they export it into a mental black box. The promise of democracy is that you can retain the *appearance* of choice, or at least voice, while offloading all the bloody work and hiding it from sight. Liberal ideology is appealing to many precisely because it avoids the responsibility of individual agency. Police, as David Graeber reminds us, are bureaucrats who constantly use crude violence to *simplify* the world.[213]

The ontology of sex and gender that Kathleen Stock advocates is not one of the bare facts of the universe, it does not cut reality at the joints; rather, her endorsement of it is a *political* move, a campaign to apply state violence in certain directions. Her constant disingenuity stems from an unwillingness to just admit this. She wants to direct the violence but not have to see the blood on her hands. She protests in an aggrieved innocent tone that she supports the "legal rights" of trans folks, only to immediately smirk that transfeminist conceptual schemas—affecting the very contours of those rights—will prove *unpopular* at the ballot box.[214]

---

213 David Graeber, "Dead zones of the imagination: on violence, bureaucracy, and interpretive labor. The 2006 Malinowski Memorial Lecture" *HAU: Journal of ethnographic theory* 2 no. 2 (2012): 105-128

214 Judith Woods, "Kathleen Stock: 'No matter what I say, to trans people I'll always be a villain'," *The Telegraph*, May 28, 2023, https://www.telegraph.co.uk/news/2023/05/28/kathleen-stock-interview-oxford-university-gender-debate/.

Questions of radical truth or social liberation be damned, at the end of the day the violence of democracy is on her side, and she welcomes it.

Convergence, Conflict, And Avoidance

In dissecting Sokal's pratfalls around the concept of sex/gender we had to distinguish between *radical* realism and *fundamentalist* realism. Both strive for simplicity, but from distinct motivations. The radical realist seeks only what simplicity actually underpins the world, whereas the fundamentalist realist seeks simplicity wholesale and immediately. Thus the radical is inclined to view the practical everyday world as wildly complex and the fundamentalist to see it as reassuringly simple. The radical sees our everyday impressions and concepts as suspect, the fundamentalist sees them as solid foundation. The radical seeks perfect compression, and the fundamentalist seeks compression that is merely "good enough," functional for their aims.

Both realists expect a certain dynamic of *convergence* in our models, but diverge in how they expect this to occur. The radical is more inclined to see convergence in something closer to individual terms, as the inevitable formula that reveals itself upon sufficient depth of investigation. Whereas the fundamentalist is more inclined to see convergence in social terms, as popular consensus, glossing over whatever few divergent crazies might be around.

Because the radical sees reality as something not easily given but to be *revealed*, there is no starting assumption that anyone born so far has yet glimpsed it; any compressive success of science has to be *demonstrated*. But the fundamentalist takes reality as something *given* and immediately accessible, so the problem becomes fighting the errors and confusions that might disrupt this knowledge.

Why do people not already agree? To the radical, this is because we are born in ignorance and knowledge requires diligent struggle to achieve. To the fundamentalist, this is a harder question. Since the Bible or phenomenology or whatever provides direct access to truth, those who deny it must be *broken* somehow. When philosophers savage her bad arguments, Stock doesn't investigate their perspectives, she expresses complete shock that the world doesn't agree with her, deciding that they must be defective, crazy, "fanatics."

This kind of fundamentalism is the strawman that the postmodernists and relativists got their traction on. So much of their literature in the Science Wars revolved around proving that two individuals or camps could disagree and be *justified* in their disagreements with one another.

> "Is it permissible for two peers to arrive at different conclusions from the same evidence? If you opt for 'yes,' then you are committed to a relativist permissivism."
>
> Martin Kusch, *Relativism in the Philosophy of Science*, 2021

But justified disagreement is no challenge to realism or to objectivism; no two individuals in all of history have ever had *exactly* the same evidence. Our childhoods and our explorations are always going to be somewhat different, even drastically different. From a core goal of compressing reality into a model does not necessarily immediately and transparently fall out the same set of heuristics or *strategies*. "Epistemic standards" are not magically *a priori* God-given principles, nor mere arbitrary social tradition, but *strategies* stochastically tested and adopted throughout our lives. Strategies that congeal on top of one another and shift about just as models do. In essence, just meta-models.

Further, it's easily demonstrable that two agents strictly following Bayesian rationalism can *diverge* when given the same evidence if their priors are concentrated in different regions. Convergence is still inevitable *upon sufficient evidence*, but there may be a very long period of divergence first.[215]

Dramatic disagreement should be the default, and yet we see an astounding amount of convergence. Folks from wildly different communities or traditions consistently discover the same models and meta-models. There is variation in perspective, but it is strongly constrained.

When one person points, in seemingly a final dramatic appeal, to the lens of a telescope and the other person to a holy book, those are not in any remote sense the actual final arbiters—remote and incommensurable, free floating and forever separate. Rather we are then obliged to evaluate further up the chain. The Cardinal believes that placing higher credence on historically transmitted claims has good epistemic results for specific reasons. And the astronomer has likewise has epistemic reasons for lower relative credence in the holy book. It may be that we can make the network of justifications for these divergent strategies explicit, but it may also be that they have become too tangled and complex to be anything other than tacit. It does not matter. The happenstance emergence of a barrier to conversation between two individuals is not a barrier to reason.

One need not necessarily be able to reason with someone else to reason internally, nor does someone's disagreement with you necessarily mean you should doubt your conclusions. Sometimes you should hesitate if the whole world disagrees with you, but sometimes you shouldn't.

The spectre of this *haunts* liberalism because it threatens to spawn uncompromising political radicals who become convinced that we *know better* and are thus willing to take unilateral direct action. We might blow up pipelines, assassinate CEOs, smuggle refugees, burn down police precincts… even

---

215 Alan Jern, Kai-min K. Chang, and Charles Kemp, "Belief polarization is not always irrational," *Psychological Review* 121, no. 2 (2014): 206–224.

ostracize Kathleen Stock. We might create a world where you're obliged to investigate the world beyond your nose and think through the consequences of your actions to a vastly wider and deeper degree—recognizing *inaction* as just another choice of action, never neutral, always complicit. This threatens not just the social order, it threatens individuals with an awareness of their ethical responsibilities.

The postmodernist tries to stop this by enshrining pluralistic diversity as the apex value. Punching a nazi is then bad because it would be suppression of that nazi's difference, his *uniqueness*. The nazi says he's not a *supremacist*, he's just trying to protect his culture against homogenizing globalism. He's just recruiting; his organization hasn't bombed any black preschools *yet*. "Diversity," in this use, means not just tolerance of differences of arbitrary happenstance or superficial culture, but ideological and moral differences, even differences on the most bare issues of fact. Thus, if the nazi smirkingly declares himself an anarchist, who are you to disagree? To assert any definition whatsoever, no matter how involved, would be totalitarian.

By contrast to the postmodernist, the fundamentalist realist tries to stop all this by enshrining reactionary populism. Punching a nazi is then bad because *it wasn't voted on*—the proper political agreement wasn't reached. After all, in the direct perceptual experience of an average person, the nazi seems like a normal guy, he isn't wearing a uniform drenched in swastikas, he says he's just a little fed up with "woke" (i.e. the moral responsibility pushed by anarchists). Any definition of fascism more involved or divergent from popular caricatures—for instance as "ultranationalism" or as a cluster concept—is then clearly wrong on the grounds of not being already obvious to the average Joe. If the antifascist researchers who've spent a decade following this nazi's entire career think they're right, then they should make their case to average citizens *via* centralized mass media, convince everyone, get legislation passed, and then call the cops. But if they take their evidence to social media or otherwise engage in horizontal communication, then they're engaged in "canceling." Even worse if they physically take matters into their own hands.

Neither variant of liberalism can afford to trust individuals with basic epistemic evaluations, to say nothing of ethical ones. The nazi may be wearing a Screwdriver t-shirt and screaming about killing immigrants, but liberalism constantly and proactively imposes *an assumption of ignorance*. An ideological declaration of the impossibility of individual knowledge.

This antipathy to individual knowledge is core to liberalism. When Toby Young's silly "Free Speech Union" legally threatened us for relaying true facts about Toby Fitzsimmons' vapid misogyny and racism, he was operating off a premise of "free speech" in which individuals should face zero repercussions for their statements. In this approach individuals have *no responsibility* to be truthful or accurate, because the very idea is seen as impossible. Individuals should rather be allowed just to throw declarations into the wind, and then leave it to society to somehow sort through these detached concepts. This was exactly Feyerabend's approach. He felt no obligation to get things right personally, only to be a provocative participant in some larger collective dialectic. He would, no doubt, find it unfair and tyrannical to hold him accountable for having championed faith healing.

Not feeling any personal responsibility and not being subject to judgement by other individuals is certainly more *relaxing* than taking on the burden of active agency, but individuals abandoning a commitment to personally get things right does not add up into society as a whole getting things right via some magical property of the ballotbox.

The notorious pseudoscientist Graham Hancock, for instance, has written quite explicitly about how he sees his job as that of a *lawyer*, to doggedly prosecute for a single position, discarding all accuracy, nuance, and context.[216] I'm sure he finds it relaxing believing that he can lie without consequences because society will clean it up.

Courts and elections are rituals and modes of social organization based on erasing or hiding the actual complexity of the relationships and distributed knowledge involved, all to relieve individuals of the weighty responsibility to think and to act for themselves.

In his essay "My Science Wars," the supposed realist Harold Fromm drew out explicit comparisons between the antirealism of the postmodernists and that of Christian Science, but he simultaneously complained of feminists as *"holier-than-thou critics, who brook no criticism because they are already in possession of the absolute truth."* Drawing from this horror of *"absolute truth,"* and in a tantrum very similar to Derrida, Fromm took a feminist letter denouncing a professor and compared such an act to the literal Holocaust. This is a fetishization of "humility" on par with any postmodernist.[217]

A consistent realist would have no trouble admitting that an individual (or disenfranchised minority) can know better than everyone else or that deep disagreements can arise among reasonable people and be functionally irresolvable through debate. A liberal cannot.

To avoid grappling with having to take a side in conflicts, liberals invariably seek the comforting shelter of some meta-rule that prescribes inaction: *"How*

---

216 Graham Hancock, "Writing about Outrageous Hypotheses and Extraordinary Possibilities: A View from The Trenches", *Graham Hancock*, January 19, 2002, https://grahamhancock.com/outrageous-hypotheses-hancock/.

217 Harold Fromm, "My Science Wars," *The Hudson Review* vol. 49, no. 4 (Winter, 1997): 599-609.

would you like it if people you disagree with likewise pursued their values?" "*If you're saying that it's acceptable to punch nazis, then you're saying it's acceptable for nazis to punch Jews; you could have no grounds to object to them doing the same.*" This, of course, does not follow. We are under no obligation to erase our actual knowledge of distinctions, just as we're under no obligation to take an absurdly humble meta-epistemic stance and assume our moral judgement of the ideology of nazis is an *arbitrary* value, interchangeable with their hatred of Jews.

Liberals believe that if you pronounce a meta-rule and bind yourself to it, deliberately erasing your knowledge, everyone else will be obliged to play along and be bound to the same.

Spoiler: It doesn't work out that way.

THE TRAP OF CIVILITY AND COMPROMISE

When not preaching before a reactionary mob, Sokal has actually admitted that terms like "female" or "woman" can be products of social dynamics in much the same way we use "parent" to be inclusive of adoptive parents like himself.

Sokal's pivot is to the gods of pluralism and compromise, a commitment not to reality, truth, or liberation but to *inclusive debate*. All his talk of his definition of sex being a "fact" is thus frankly admitted to be dishonest posturing.

But it's justified, you see, because the other side is refusing to *compromise*.

> "They want to demonstrate that there is nothing to debate—that a thorny social and political issue is in fact nonexistent. No need, therefore, to give sensitive and empathetic consideration to the legitimate—and unfortunately conflicting—interests of different groups of people. No need to discuss respectfully across identity and ideological lines, and to craft fair compromises."
>
> Alan Sokal, *The Bad Faith Use Of Words*, 2022

Now don't get me wrong, I think nuance and clarity is important and it can be valorous to engage with the distinct linguistic—or even political—frames of one's interlocutors in honest and open ways. But there is no moral obligation to *compromise* on matters of liberation. It should make not one lick of difference if the majority of voters are monsters. An attitude of deference to democracy would never have smuggled guns to slave revolts or hidden Jews.

In the same piece, Sokal whines about the uses of "racist" and "fascist" being definitionally laden, even links critically to an IllWill[218] article. And it's certainly the case that our definitions of sex/gender, racism, and fascism are directed by both a political analysis and value system, just as any alternatives will be. But should we *compromise* with those whites who complain that the erosion of white supremacy has eroded their power? Should we give ground on their linguistic definition of "white genocide"? Should we call that a *"thorny social and political issue"*? Or should we just meet them with baseball bats in the street, as with the French antifascists whose actions so horrified Latour?

In this, Sokal follows his compatriot Chomsky, who infamously embraced and defended the Holocaust denier Robert Faurisson on the grounds that defending basic reality was of far less importance than assuring a nazi never experienced any negative social consequences for lying.

218 Does a has-been radical leftist like Sokal have thoughts on the various anarchist versus leftcom fights around the resurgence of tiqqunism, projects like Inhabit, and the racial politics of *How It Might Should Be Done*? Or has his deeply uninquisitive ass only discovered IllWill through some circle of Mumsnet reactionaries hyperventilating about Judith Butler? You can't help but want to know.

It is beyond the scope of this book to repeat the entirety of the case for refusing to share a debate platform with fascists that antifascists have come to consensus on over decades of struggle. I've summarized the rich justification and insights of antifascists at length in other spaces.[219][220] But suffice to say the Liberal fetish for civil debate and compromise is notoriously poorly thought out and regularly exploited by reactionaries. It combines an impoverished model of how social change occurs with an artificially flat approach to evaluating information and social connections. It strips out content and context and calls that a virtue.

It's darkly humorous that Bricmont and Sokal, in particular, got so thoroughly gamed by batshit reactionaries, because physics is no stranger to these maneuvers. For every crank who has a Theory Of Everything that is being *censored!!* by physicists not answering his emails, there's a hundred scoffing that modern physics is a tower of garbage because one or another conclusion or theoretical program isn't intuitive to them. And always the same vapid appeal to debate and civility: *"They won't debate me because they're afraid! They laughed at me which is uncivil and an ad hominem!"*

The very coiner of the phrase "Intellectual Dark Web," Eric Weinstein, spent years hyping up his theory of everything, Geometric Unity, while refusing to give details or formulas. When he finally released it and got sharp criticism from physicists, his reaction was furious name calling, dismissal, and silly demands. Stock and Lindsay are basically no different. They focus their efforts on gaming the referee of liberalism, sucking up to newspapers or YouTube clicks.

It's a simple formula, visible with countless far-right influencers online: you construct lazy defenses of populist baby assumptions, scream that you're being censored because no one is legitimizing you with debate, then fall into a fainting chair and act like you've been murdered when you're critiqued.[221][222]

The core objection of the "grievance studies" trolls was that in many academic papers it's taken for granted that things like patriarchy and white supremacy are institutionalized in our society. As Lindsay griped, *"It's not did*

---

219 William Gillis, "Responding to Fascist Organizing" *Center for a Stateless Society,* January 25, 2017, https://c4ss.org/content/47734.

220 William Gillis, "Antifa Activists As The Truest Defenders Of Free Speech" *Center for a Stateless Society,* November 19, 2017, https://c4ss.org/content/50151

221 Talia Mae Bettcher, "When Tables Speak": On the Existence of Trans Philosophy," *Daily Nous,* May 30, 2018. https://dailynous.com/2018/05/30/tables-speak-existence-trans-philosophy-guest-talia-mae-bettcher/

222 Kim Hipwell, "The Matter Of "Material Girls"," March 12, 2023, https://kim-hipwell.medium.com/the-matter-of-material-girls-71bf7a488fbe.

racism occur, it's how did it manifest in the situation." [223]

This is the same as objecting to studies on how global warming is impacting local climate in a region because it takes the existence of global warming for granted, or griping that physics journals print results of electron scattering without arguing for the existence of electrons in each issue. It's a deeply anti-intellectual and populist demand that every discussion or idea be no more than baby's first introduction. Such an expectation would forbid compounding theoretical depth, obliging everyone to do nothing but eternally revisit and reaffirm the most basic analysis before hordes of flat earthers and 13-year-old suburban white boys.

We can surely do more to streamline introductions to ideas in physics or anarchism, but mere *inaccessibility* to random normies is not an invalidation, nor should advanced work be suspended to forever defend the most basic facts.

The inconsistency of this demand exposes the dishonesty and disingenuousness of it, as it is only applied to implicitly assert the relevance of one set of social circles versus another. Lindsay is pickled in conservative hogwash about white supremacy being over and a "white genocide" looming, so he is outraged that we're not Teaching The Controversy. Bricmont is pickled in Holocaust deniers, so he does the same. Some old gender reactionaries were nice to Sokal, so he embraced them. Lindsay heaped praise on Sokal and seemed to make him relevant again, so he embraced Lindsay. Those outside their social circles have strange ways that must be obviously without merit. And if they do not engage in displays of deference towards you and your friends, studiously walking you through an introductory course while taking your barbs without complaint? Well in the face of such incivility there's no need to investigate. No need to operate in good faith. No need to continue admitting there's no such thing as a "fact" of a single definition of "sex," when you can just regurgitate applause lines before junior tories.

Sokal is fond of referencing John Stuart Mill on the benefits of engaging with opposing ideas—amusing given how little he engages with the feminists and antifascists he denounces—if only to better understand your own arguments. But investigation and engagement with *arguments* or *ideas* in no sense obliges civil public debate with *people*. Feminists constantly write devastating takedowns of the arguments of transphobes like Stock, but that doesn't mean they should treat her like a legitimate discursive rival, any more than biologists should treat racist pseudoscience as worthy of debate in journals.

---

223 James Lindsay, "Applied Postmodernism: How "Idea Laundering" is Crippling American Universities," talk at Aspen Jewish Community Center, July 30, 2019, *GrassRoots Community Network*, uploaded August 9, 2019, YouTube, 44:45. https://www.youtube.com/watch?v=XeXfV0tAxtE.

For all the *"they don't even know what they're talking about, they just call everyone a nazi!"* nonsense that saturates conservative media, antifascist researchers are not only more fastidious about epistemic diligence and details than anyone in the media, they have long been experts in the various arguments and views of the various branches of the fascist movement they surveil, infiltrate, and monitor. When you or a comrade will be immediately murdered if a cover is broken, there's some serious incentive to get epistemics and finicky details right. They build such knowledge often with greater breadth and depth than many of the fascists regurgitating such, all while reactionaries regularly fumble at the most basic understandings of antifascists (or don't bother trying to get anything right when lies can get them acclaim). Antifascists have achieved such deep and accurate knowledge of the enemy for decades *without* legitimizing them with the prestige and reach of a platform.

One can argue that the US media's decision to nonstop platform fascists since Trump came down that escalator changes the calculus a little—or the way the entire media establishment in the UK has collaborated in aggressively pushing rancid transphobia—but the point is that accurate models do not depend upon debate.

Beyond the way that discursive engagement with rabidly bad faith adversaries provides them with benefits, debating does not always strengthen one's own understanding of the truth. We all know the experience of engaging with someone of sufficiently inane, confused, and inept—to say nothing of morally horrific—positions and coming away diminished, fatigued, or even scarred for the experience. We owe nothing to those operating in transparent bad faith like Stock, nor do we owe anything to people who simply do not share our most basic moral and political values.

In short, if we're serious about reaching truth in social epistemology, there are serious exceptions and complications to the prescription of open debate.

I am a passionate militant hardliner about free speech, but *actual free speech*, like leaking government secrets, pirating "intellectual property," and advocating in detail for the royal family of the British empire to be assassinated. Not "civility."

Is free speech about the free flow of information, a consequentialist tool in pursuit of truth? The sort of free speech that antifascists clearly leverage when they expose the words, actions, and identifying information of fascists. Or is it merely a commitment to being polite? As the "Free Speech Union" clearly meant when they threatened us for exposing Fitzsimmons' participation in misogyny and racism on the grounds that sharing the truth constituted "bullying."

Sokal clearly thinks the latter, because he has repeatedly cited and praised a piece by two former feminists that argue it explicitly:

> "It should be a basic right for all workers to take part in the democratic process without fear of losing their livelihoods."
>
> Alice Sullivan and Judith Suissa,
> *The Gender Wars, Academic Freedom and Education*, 2021

I would hope anyone reading would recognize that's just *obviously* a horrifically evil prescription. If someone calls for the genocide of a racial minority, it's an infringement upon freedom of association to demand that individuals continue to work with that individual. I know of quite tangible examples where a worker at a democratic cooperative went off the rails into things like Holocaust denial or rabid misogyny. A supposed "right" of said worker not to have to find another income source is clearly trumped by the right of their fellow workers not to deal with them.

The entire fucking point of freedom of speech is to get information to individuals so they can make informed choices. It's about expanding agency. To deny someone the choice about whether even to work with someone is to make a mockery of this. Your co-worker is allowed to talk endlessly about how much he wants to butcher and eat you, but you're not allowed to tell him to fuck off. Suissa and Sullivan's approach is a deeply authoritarian one because it obliges state power intervening to secure a worker's "right" to associate with you against your freedom of association.

Additionally, such an approach forces all activism into the state nexus. It would undermine the most basic category of anti-authoritarian organizing outside the state, like boycotts dynamically catalyzed from the bottom-up.

Just as the free speech of gossip circles have empowered women and other oppressed minorities to organize resistance arguably since the dawn of patriarchy, stateless societies are usually predicated upon similar. In game theoretic terms, altruistic minorities can out-punch the selfish or apathetic by leveraging asymmetries that benefit the self-sacrificing. Just as a single civil rights activist with a gun can hold off dozens of Klan members unwilling to be the first to die, boycotts can be amazingly effective at leveraging the twin foundations of stateless societies—freedom of association and freedom of information—to bypass centralized power and suppress metastasizing power. Suissa and Sullivan's approach attacks the very foundation of freedom.

Association matters. Just as it was discrediting of *Social Text* in 1996 that they chose to publish the crystal clear batshit antirealism of the Hoax, it's discrediting of Sokal that today he chooses to associate with and praise vapid reactionary grifters.

Anyway, I've detailed antifascist arguments and drilled into the anarchist rejection of democracy,[224] I've detailed at length the inherent and stark abusiveness of opponents to "canceling."[225] Just as there's inherent tradeoffs between civility and truth, there's inherent tradeoffs between civility and liberty.

---

224 William Gillis, "The Abolition of Rulership or the Rule of All over All?" *Human Iterations,* June 12, 2017, https://humaniterations.net/2017/06/12/the-abolition-of-rulership-or-the-rule-of-all-over-all/.

225 William Gillis, "A Giant Red Flag Folded Into a Book" *Human Iterations,* December 6, 2022, https://humaniterations.net/2022/12/6/giant-red-flag.

REALISTS ARE THOSE WHO SHOVE BACK

One thing that I'll grant to both Sokal and Bricmont is that even while they clung to the liberal ideology of democracy, they were socialist enough to at points recognize the infamous tepidness of liberals in social struggles.

> "Denouncing and stopping are two very different things."
> Jean Bricmont, *Humanitarian Imperialism*, 2006

When Jean Bricmont complained about liberals echoing empty words they did not truly believe—merely performatively wiping their hands of US imperialism rather than struggling against it—he was not wrong.

There is no such thing as neutrality. Every action (or inaction) we take is embedded in a wider causal universe. There is no getting out of this entanglement, no option of running from reality. Further, while science strives for universal lossless compression, every conceptual structure we make at the messy human scale is unavoidably perspectival and directed. In such matters we face overwhelming computational complexities that cannot be compressed without a loss, without bringing some things into focus and obscuring other things.

Physical reality provides strong *boundary constraints* to our models of it at such scales, but there is always the question of how much to gloss over in our description of it, how parochial or universal to make our accounts.

Bricmont and Stock have reactionary inclinations: to stick with the simple, regardless of how lossy. Bricmont reduces geopolitics to a simple tale of conflicting armies where all enemies of one's enemy are automatically collapsed into comrades. Stock reduces patriarchy to a simple tale of conflicting armies, where the only thing that really matters is genetics, and the only solution is perpetual statism.

> "Political problems are generally not intellectually complicated."
> Jean Bricmont, *Humanitarian Imperialism*, 2006

But just as overcomplicated descriptions can be used to obscure simple realities, so too can overly simple accounts obscure important mechanisms and possibilities. Because reactionaries hunger for simplicity and only care about *"good enough"* in their models they inevitably reinscribe the existing order, never seeing past its horizons.

Radicals strive for more universal accounts—perspectives that can commute between different contexts, the view that can persist *anywhere*. But such universality is not even remotely the same thing as neutrality. As we traverse reality, it reaches into us and reshapes us. Exploration may leave us less attached to our arbitrary origins, yes, but we converge into new perspectives.

If the universe were makeshift, incoherent, or chaotic, there would be no such pressure. A willingness to explore would be synonymous with disconnected neutrality. This is the nihilism that undergirds liberalism, where intellectual seriousness is seen as necessarily disinterested and passive.

When the incipient anarchism of the Diggers was crushed in the English Civil War, modern liberalism emerged in the bloody wake as an attempted *compromise* in the eternal war between freedom and power, an attempt to freeze the battle lines in place via a framework of civil and disinterested consensus making. The Royal Society emerged out of this context. Just as realism was frowned upon and instrumentalism encouraged, what marked you as lordly or reputable was *not having a stake*. And this class culture passed from the aristocracy to the bourgeoisie.

What liberalism promises is that if you epistemically mutilate yourself down to sufficient humility, you will no longer be obliged to act, your passivity will grant you a place of prestige above conflicts and moral appeals. When former atheists like Richard Dawkins share stages with authoritarian christian nutjobs or racist pseudoscientists they all congratulate themselves on their civility, which becomes magically transmuted into a badge of their objectivity. They are not passionate. They are to be taken seriously as studious seekers of truth *precisely because* they are collegiate with monsters.

> "'Civility is racist' was a central principle of the Evergreen revolutionaries and their faculty mentors, even if they never said it in exactly those words."
> Brett Weinstein, Twitter, 2024 [226]

But just as *"teaching the controversy"* of creationism or putting both a flat earther and a physicist on debate platform are not neutral, a demand for civility between a slave and a slave owner is not neutral; it's participation in the enslavement.

In the aftermath of his hoax, Sokal was primarily denounced by *Social Text*'s wider circles for having breached standards of civility. He had violated the norms of academia and stepped outside his proper role as a scientist. He had treated ideas like they matter, like the game was *real*. This was his ghastly crime.

In the liberal imaginary the ideal of scientific objectivity is thus not a matter of *the world* having the firmness of an object, but of the *scientist herself* being an immobile well-behaved object, devoid of desire or rage.

---

[226] https://archive.is/NCgMJ.

The only question for the liberal is whether science should be seen as a legal system—a neutral background referee, an unimpeachable authoritative framework beyond ideology… or whether to view science as a civil service—with scientists as instrumental peons, taking orders and serving as janitors and stewards for the interests of some democratic community.

The liberal technocrat believes science should be placed *above* the fray of ideological combat whereas the liberal pragmatist believes science should be placed *below* the fray.

What the liberal cannot imagine is the scientist as an active combatant *in* the fray, a radical ready and eager for the fight.

The liberal can allow that global warming is "real" but not so real as to oblige blowing up oil pipelines. The liberal can allow that a religion that denies physical reality is "false" and getting children killed, but not so false that you grab some friends and smash up their reading rooms.

At the end of the day, the liberal cannot allow for anything to be so real

CONCLUSION

"There may, of course, come a time when it will be necessary to give reason a temporary advantage and when it will be wise to defend its rules to the exclusion of everything else. I do not think we are living in such a time today."
                                    Paul Feyerabend, *Against Method*, 1975

"We seem to be hurtling towards a general reconciliation. But perhaps it is not too late to draw back from the brink of peace."
                                    Steve Weinberg, *Peace At Last?*, 2001

"Peace in chains is an affront that should be refused. There is peace in the dungeon; there is peace in the cemetery; there is peace in the convent. But this peace is not life... Let such a peace be damned!"
                                    Ricardo Flores Magon, *Preaching Peace is a Crime*, 1910

Let us summarize the Science Wars:

In the wake of the second World War, the military industrial complex purged the formerly radical-infested sciences while much of the aristocratic thought that had characterized humanities found continuation among a glut of students. Thus were the political aesthetics of the humanities and natural sciences reversed.

At the same time, a shocking misdiagnosis of fascism as universalism gained popularity and led to a subcurrent of the Left feeling there was a moral obligation to support any belief that wasn't hegemonic.

These academic and subcultural currents conjoined in Paris and New York, around the Sorbonne and Columbia University, into an upstart ideology that cyclically reinforced countercultural mysticism with academic prestige.

Thomas Kuhn and Paul Feyerabend achieved fame by aggressively playing to this current, but each eventually became disenchanted and frightened by it—although in Feyerabend's case it seemed to take his terminal cancer to partially reverse course. Others who had embraced this audience, like Richard Rorty, never broke ranks, but made increasingly explicit their motivation from authoritarian (i.e. liberal) commitments directly opposed to anarchism.

Amidst all this, Stanley Aronowitz sought to leverage these relativist and antirealist currents to defend marxism by attacking the reductionist project at the core of science. This drew the attention of a pair of reactionaries (Levit and Gross) whose populist book dragged many antirealist claims circulating in postmodernist-influenced academic circles into the sunlight.

This attention spurred the young socialist Alan Sokal to read up on postmodernist literature and write a takedown of Aronowitz's arguments, rhetorical devices, and willing association with hardline antirealists. When this was submitted in the thin guise of a fawning fan letter, Aronowitz and his friends blithely published it. In the fallout, the academic and subcultural sides of postmodernism defensively closed ranks, trying a variety of disingenuous rhetorical and narrative strategies. And the relativist sociologists who'd caught some strays jumped in and got repositioned as the respectable face of antirealism, while Aronowitz continued praising, publishing, and associating with countercultural mystics like Lamborn Wilson, even as the latter defended chaos magick and child rape.

The war created dramatic conflicts within the Left that continue to this day, but in the public's mind it was an intra-academic affair that died a slow death as realist physicists and relativist sociologists fizzled out their hostilities in tepid debates and conferences.

Younger generations of continental philosophers were embarrassed by their milieu's embrace of linguistic idealism and turned to things like critical realism or speculative realism. By 2004, and Bruno Latour's public *mea culpa*, neither postmodernism nor criticism of it was all that interesting to academia.

But *outside* of academia significant currents of the Left and the Right enthusiastically embraced the relativist and antirealist arguments that had been justified on political grounds. Edgy mysticism or skepticism remained cool in leftist subcultural spaces and extreme relativism propagated among those committed to nationalism-of-the-oppressed. Meanwhile, conservatives directly copied postmodernist critiques of science to push creationism, climate denial, and conspiracy theories, while fascists recognized postmodernism as a continuation of their defense of national particularity against universalism.

At the same time, mass-market conservatism never stopped using *"they don't believe in truth because it's on our side"* as a readymade explanation for the existence of any argument out of the Left that conservatives were too intellectually lazy or reactive to grasp, and eventually Jordan Peterson would revive this tradition by loudly blaming trans people and all egalitarian values on "postmodernism."

A few reactionary grifters attempted their own much lazier hoaxes, this time targeted at minority studies. At which point crusty dinosaurs like Sokal, who had started breaking bread with reactionaries to prove they were good liberals, felt the wind behind their tattered sails again and embraced the fascist hordes.

The postmodernist holdouts found themselves abruptly revived by Peterson's hate, and they have continued the same old motte-and-bailey games, enthusiastically embracing mystical subcultural practices as nebulous *epistemic resistance* while retreating under any pushback to various shoddy antirealist philosophical arguments.

This book has attempted to trace all this tangled discursive history to expose everything that was hidden, the possibilities abandoned or lost. We do not have to choose between relativist ethnonationalism, idealist mysticism, expert technocracy, and conservative fundamentalism. Within the traditions of both physics and anarchism there is an alternative: a *radical realism*.

Turning Their Tools Against Them

Just as I have turned the sharp steel of realist reductionism against the reactionary politics of Sokal, so have I ruthlessly turned every rhetorical tool of the postmodernists against them. From evocative poetics and personal experience, to emphasizing institutional incentives and associations, I've adopted all their favorite maneuvers to flesh in a competing grand metanarrative that I know will feel starkly alien—even insane—to those raised in the discursive remains of postmodernism.[227]

For many in the humanities there is not the faintest notion that there could be perspectives native to the natural sciences with greater claim to liberatory politics. In conjunction with the institutional and class dynamics of academia, this has cultivated a discursive asymmetry whereby all sorts of intense moral and political attacks can be lobbed daily by the humanities against the sciences (*everyone knows* that rapacious behavior by tech CEOs would somehow be solved if STEM nerds took more poetry classes), but to reverse these attacks is unthinkable aggression and scandalously unfair.

That's part of why this book has taken such a sweeping and comprehensive approach to discursive history. To tackle any subsection of things would leave the overall metanarrative of postmodernism room to retreat or ignore. Any one of the arguments or accounts I've given in prior chapters deserves multiple book-length treatments, but, in my experience, when you press any one point or critique you can easily convince folks, but they continue seeing that as merely *an interesting exception* to their wider metanarrative.

When Steve Fuller, for example, notes that the sheer volume of content defending an existing order can suppress dissent, *"heavy documentation interpreted in one's favor creates a burden of proof against any attempt to challenge the legitimacy of one's rule,"*[228] he's trivially correct. But this is pretty much the entirety of how the postmodernist metanarrative has justified itself: with just piles and piles of content. Too much to be engaged with quickly or succinctly, so the whole beast lumbers on.

> "When one tries to follow up on its frequent allusions to what has been 'established elsewhere,' this usually means one of the following: theoreticians interpreted in the light of other theoreticians; debatable hypothesis made the a priori premises of arguments; jargon employed to give cliches and commonplaces an air of profundity; interpretations which, through question-begging terminology, presuppose what is being demonstrated... A similar gimmick

---

227 "*Demonic*" is an epithet once frantically spat about me by a certain postmodernist and tiqqunist.
228 Steve Fuller, *Science: Concepts In Social Science* (2007), 98.

> might be called the phantom citation. Postmodernist literary scholars will write whole books just by announcing that so and so has put X 'into question.'"
>
> James Drake, *The Academic Brand of Aphasia: Where Postmodernism and the Science Wars Came From*, 2002

We all know that academia has been structured into an engine for producing piles and piles of content. And just as this incentivizes *p*-hacked studies in the less radical sciences, in the humanities this can produce papers of little value or insight. While reactionary grifters like Boghossian and Lindsay were motivated by opposition to the egalitarian values and liberatory politics claimed by minority studies, we all know that academic fields incentivize the publication of torrents of vapid slop.

Thus, while few academics will ever burn down a police precinct, many treat radical political movements as a resource to be mined for aesthetic cachet and insights that don't have to be attributed. There's every incentive to gesture at explosive claims, so as to draw support from movements outside of academia and to bolster your prestige within it.

> "Much of the antagonism of the science wars can be traced to the well established academic practice of stating one's views in extreme form to stir up a controversy and thereby attract the kind of attention that can actually enhance a career. This is common practice in the humanities and not unknown in the sciences."
>
> Craig McConnell and Robert H. March, *Bringing Reason and Context to the Science Wars*, 2001

Academics want to pretend that they're fresh, underground, and outside power—a characteristic book like *Underground Theory* presents contributors on its cover like the lineup at a concert—while simultaneously being a wealthy greying establishment, utterly complicit in our hellworld. Most academics are simply playing a different game than activists, trying to sound more extreme than radicals, while preserving their insulation from any stakes *via* appeals to authority and civility. Thus they usually prepare lines of retreat through indirect reference or textual obscurantism.

But while some academics may not have been that committed to their antirealist flirtations, when they position themselves as the only source of intellectual or theoretical authority one cannot blame activists and subcultural leftists for taking the stakes and claims seriously.

Students enter academia, look for ostensibly egalitarian or liberatory voices, and find a massive pile of literature that agrees in broad strokes, reinforcing a common metanarrative that still inherits much from postmodernism and the critique of universality as totalitarianism. I've lost count of the times some rich kid from a private college has confidently lectured to me about how *everyone knows* that "science" genocided all the witches in the service to white supremacist patriarchy. When I lightly push back and bring up my status as a disabled survivor of a faith healing cult, they are inevitably drawn up short and gripe that to play such a card is "unfair" and "identity politics."

There is no pressure upon the comfortable to be aware of dynamics that daily benefit them. Folks will press you in a vice for decades, but the second you place them in that same vice they scream bloody murder, without a second's thought for how you so intimately came to learn its function. I will admit to having cultivated a passing sadism on this front. But I have studiously restrained from simply returning every tit-for-tat.

The increasingly dominant account of the science wars is just *false* in many concrete respects. The point of Sokal's hoax had nothing to do with peer review; he supported *Social Text*'s role as a marxist zine and sociological criticism of science as an institution. There were, in fact, many people in postmodernist circles making quite strong denials of physical reality. Postmodernism was neither the origin nor the vanguard of the radical activist traditions now so influential, but often their enemy. And both postmodernism and the relativist sociologists were broadly influential upon the far-right, directly furthering climate denial and setting the stage for the "post-truth" resurgence of fascism we now find ourselves in.

One can, of course, attempt to contest this broad narrative to some extent by denying that "postmodernism" is a descriptively useful coarse-graining. Labels and emphases can be disputed. But the connections and dynamics that I have mapped survive. As does the potential resonance of the alternative valences and overall narrative I've sketched.

At the very least I hope to have punctured the absurd attempts by liberals like Rorty and Feyerabend to frame their antirealisms as speaking on behalf of anarchism and anti-authoritarianism.

THE REACTIONARY CONJUNCTION OF
SKEPTICISM AND FUNDAMENTALISM

It almost goes without saying that the mainstream conservative narrative on the Science Wars is likewise trash, but let's reiterate that Jordan Peterson blames the postmodernists for the very things they were in open conflict with.

Lamborn Wilson was a staunch enemy of both antifascism and feminism. Even in the issue of *Social Text* that tried to rally the wagons, it was admitted that there'd been sharp tensions between postmodernists and feminists. Almost all the academic postmodernists screamed bloody murder about the need to maintain civility, and I know many still who gripe that antifa are "the real fascists." Even before Lotringer's conference was derailed by feminists and anarchists, postmodernism had inherited the critique of anarchism's moral universalism and decentralized resistance as "totalitarian." For those of us anarchists, antifascists, and feminists who created everything conservatives are now apoplectic about—from deplatforming nazis to cancelling rapists—the attribution of these tactics and successes to those in the Left *most opposed to them* is beyond galling.

An allegiance to epistemic rationality, objective truth, and physical reality is not at all the same thing as a commitment to civility and persuasion at the exclusion of all other strategies. It is *precisely* because anarchists, antifascists, and feminists know that the world is real and the stakes are all too real that we recognize the value in decentralized boycotts and violent resistance. It is the conservative ideology of liberal democracy that functionally treats reality as exclusively socially and discursively constituted, that demands we ignore basic facts and tie our hands to preserve the illusions of the state.

Conservative grifters worship the crudest lowest-common-denominator image of "debate" because they consistently lose on any topic not sliced down to shallow and immediately accessible soundbites. What we have long referred to as "sex" is simply not binary at root, but a far more complicated ecosystem of dynamics, and to blithely assert the binary we first percieved as fact requires not just conceding to Aronowitz's embrace of teleology but abandoning the reductionist epistemic virtues of science for something much more lossy and flagrantly perspectival (in the service of patriarchy), a worship of the merely *simple* rather than *compressive*.

Conservatives like Peterson are thus invariably pressed into embracing a host of positions from postmodernism. But all the same, they *need* to accuse us of being antirealists to explain why their simplistic pictures of the world—from accounts of the necessity of social hierarchy to race pseudoscience—have not flourished but increasingly been rejected. For them, if it isn't populist common sense and reassuring defense of the status quo, it cannot be "truth."

Reactionaries are, at base, nihilistic skeptics, who do not believe in convergence across perspectives, and are thus terrified of leaving home lest they become unmoored and adrift forever. Theirs is not a radical allegiance to truth or reality—to be painstakingly discovered in all its strange and alien features—but an allegiance to convention, intuition, sensation, and starting assumption. They believe in an arbitrary relativistic world where "truth" is parochial loyalty.

This is how they are always simultaneously extreme skeptics and extreme fundamentalists.

The history of the Left after the second World War has unfortunately been characterized by a forgetting of this. The misrepresentation of fascism as "totalitarianism" stripped away awareness of its core skeptical and particularist dimensions, allowing reactionary arguments and perspectives with an undeniable fascist heritage to take root in the Left, before—inevitably—getting imported back to conservatism to supercharge it in the post-truth era.

But one thing that conservative critics did not make up is that the claim that *there is no alternative to power and all linguistic interactions are games* permitted widespread conscious dishonesty among postmodernists, just as it today empowers conservative conspiracy nuts and fascist streetfighters.

The most arresting discovery I had—as a poor kid who'd already been organizing and fighting the cops for years—upon entering a private liberal arts college was the open smirking pride that postmodernists would take in sophistic maneuvers. The smug twinkle in their eye when they'd make a rhetorical move that was flagrantly unfair but would take far more effort to formally hold to account. They would gesture at the grandest moral stakes, while clearly wedded to nothing other than their own privilege.

MATERIAL AGENCY AND CLASS IDEOLOGY

> "An academic will help the cops arrest you while sneering that if you don't think a rock is a person you're reinscribing western colonialism."
>
> Mia Wong, Twitter, 2022 [229]

I've focused on postmodernist attacks against the radicalism and universalism of physics because I care about the strongest defenses of physical realism, but such attacks were obviously part of a broader disinterest in materiality that included the most pragmatic and brute particulars too.

There simply is no need to carve reality at the joints when you're not actually interested in taking knives to the existing order. Postmodernists will happily talk for hours about feces—as an airy bundle of Bataillean poetics—but when it comes to getting the community center's toilets to work, that's left to the rest of us.

> "I myself find inescapable the suspicion that strong constructivist and relativist positions embody what seems patently an ideology of the powerful. Only the most powerful, the most successful in achieving control over their world, could imagine that the world can be constructed as they choose, either as participants or as observers. Any who lack such power, or who lack an investment in believing they have such power, are painfully aware that they negotiate an intransigent reality that impinges on their lives at every turn."
>
> Alison Wylie, *The Interplay of Evidential Constraints and Political Interests*, 1992

The materiality of class is *essential* to any understanding of postmodernism as a movement, milieu, and ideological project.

To admit this does not require us to engage in the class-reductionism of so many marxists, who see struggles for trans liberation or any concern beyond some abstract white male worker in a hardhat as distractions from "materialism." The ideological, cultural, and interpersonal dynamics of cisheteropatriarchy and transmisogyny, for example, are not dismissable epiphenomena floating on top of some economic, technological, or environmental level of analysis where everything 'real' lives. Nor is liberation a matter of putting everything else on hold to further the unity and victory of "the working class" at the expense of everyone labeled "lumpen."

An important distinction must be made between the attempted radicalism of marxism and the *actual* radicalism of anarchism. Marxism arose from an analysis of society that claimed it was rooted in some "material" layer (in

---

[229] Unfortunately lost to Mia deleting her account and it being missed by the Internet Archive.

the messy abstract level of economics, not down to the bones of physics), while requiring laughable collapses of historical context[230] and necessitating awkward epicycles to make sense of white supremacy, patriarchy, ableism, etc. Anarchists are derided for the simplicity of our core analysis (the ethical critique of *all* domination, all constraint of agency) because this tight core immediately unfolds into a vastly complicated strategic analysis and entangled account of the world that can't be collapsed to a simple script like *"the commodity form and means of production determine everything else"* or *"unify the working class, then seize the state, whereupon a miracle will occur."* But this disagreement over radical analysis does not imply that we must, like liberals, at best reduce class into one more interchangeable component in a soup of oppressed identities.

Class is not merely a matter of identity, in much the same way that sex/gender is not merely a matter of identity, but something both complicated and entangled with all-too-firm material dynamics.

Just as sex/gender is not entirely free-floating from the material constitution of our bodies—the hormones they run and the structures they form, as well as the dangers or opportunities they open us to in the desires or ends of others—material concerns and constraints shape class to an immense degree. Poverty has a distinct bite that is always important and should never be glossed over. There is a *hardness* to something not being physically possible with the tools or resources in your reach. Other lines of oppression are not reducible to it, but any evaluation of patriarchy or white supremacy that makes no reference to the material is impoverished in the extreme.

When radicals repeatedly emphasize that postmodernism is a construction of rich art kids slumming it and comfortable professors wedded to academic institutions, that is not to strip away any other identities or experiences of oppression they might have. Class isn't everything, not even close. But there's systematic impoverishment in the resulting analyses that's immediately and overwhelmingly visible to anyone from poverty. Say what you will about college marxists playacting at salting workplaces or trustfund oogles whipping out their credit cards whenever they get tired of dumpstering, but postmodernism is the one current in the Left that is almost *entirely* captured by those with a history of insulation from real poverty.

For decades trans femmes have critiqued postmodernists for mining them for abstract questions on gender that can then be used by middle class academics—who are overwhelmingly not trans femmes, and certainly in no economic situation like homelessness—to spin theories, while ignoring incredibly

---

[230] William Gillis, "Anti-Engels (or Anti-Anti-Duhring Aktion)," *Human Iterations*, November 5, 2021, https://humaniterations.net/2021/11/5/anti-engels.

important dimensions around economic coercion and class:

> "The epistemological shortcomings of certain forms of feminist theory are causally related to their methodological choices... Because Butler has not engaged in a detailed, careful study of a milieu, the theoretical and political frames she proposes are equally insufficient, based as they are on incomplete information."
>
> Viviane Namaste, *Undoing Theory*, 2009

This arises from a *false generality*, whereby notably precarious trans femmes are simultaneously used as examples while subsumed and erased into far more sweeping categories of gender, so that the specific oppression faced by an HIV-positive homeless black trans femme street walker can be claimed by "trans people" as a whole. What is repeatedly—instinctively—stripped away by these academics is the *materiality*, raw structures of reality that are not directly socially fungible.

An academic might face severe dynamics around status, voice, and access, but these are attenuated from the physicality of survival sex to get someone to watch your back in the camp by the airport. The tradeoffs made between the clothing required for clients and the bite of the winter air. Gender is performance, sure—*of course*—but the liberal middle-class mind prefers to see *only* performance. It prefers not to delve too deep into the mechanics of physical violence at play or what will get the car working before the cops arrive to clear the camp.

Just as the workers walking past a tiny child in an oversized hoodie huddled against a brick wall wanted to ignore my existence, it's *uncouth* to discuss issues of raw materiality too directly in academia. There's a strict class code, and it involves assuming no one present is a *normal person*, that is to say: familiar with the way hunger and malnutrition make your joints hurt.

In this context, postmodernism's obsession with open interpretative networks and complex hermeneutics is an astonishingly explicit expression of academia's class interests and material complicities.

> "[In] the birth of critical resistance with funding from non-profits, to embedding abolitionism in the academy, you are drifting away from the ground, the terrain, of suffering and sorrow and rebellion... And so what we get is some derivative of a desire of being free that we no longer can articulate because our language itself belongs in the academy. Our imagination itself belongs to the academy."
>
> Joy James, *The Architects of Abolitionism*, 2019 [231]

---

[231] Joy James, "The Architects of Abolitionism," The Architects of Abolitionism: George Jackson, Angela Davis, and the Deradicalization of Prison Struggles lecture, April 8, 2019, *Brown University*, uploaded May 7, 2019. YouTube, 39:35

Academia can excitedly assimilate almost everything; it can embrace queerness, race, gender, disability, etc., any *difference* when treated as a purely social matter, but it cannot come to terms with the often brute painful simplicity of materiality. To do so would lay bare how the exclusionary, hierarchical, and exploitative institutions of academia secure the relative insulation of a chosen few—yes, even adjunct professors and students—at the cost of violence elsewhere.

In this context, to admit the pressing relevance of exactly how matter is distributed or arranged would be to admit that much of what "left" academia has spent decades producing at such cost is not tangibly liberating anyone.

> "Trans women need, more or less in order: decriminalization, housing, education and employment. As in, not being swept off the street, not being banned from shelters, yes being allowed in GED classes, and, well, employment. Can you provide these? Not as a goddamn researcher, and probably not as a member in good standing of whatever professional body you aspire to join."
>
> Anne Tagonist, *Fuck You and Fuck Your Fucking Thesis*, 2009

https://www.youtube.com/watch?v=z9rvRsWKDx0

War Without Allies

In *Prophets Facing Backward*, Meera Nanda argued that marxism inadvertently played a huge role in facilitating rise of postmodernism in India—and the resulting boost to hindufascists—because an inane reading of materialism convinced that directly arguing with and critiquing postmodernism would be of no effect. Because ideology was seen as directly determined by material conditions, many in the People's Science Movements chose to shut up and avoid pushing back on their "allies" in the Left.[232]

Again and again, this push for mass, solidarity, and civility is how liberatory movements lose their way and end up empowering reaction.

Judith Butler was right when they pointed out that "woman" and "man" unavoidably function as cluster concepts heavily shaped by performance, identity, and institutions. And they were wrong when they sided with the sexual predator Avita Ronell.

Alan Sokal was right when he tore into the antirealist currents metastasizing in postmodernism. And he was wrong when he sided with reactionary grifters like Lindsay and Stock.

All throughout the Science Wars, the worst shit was upheld by an instinctual campism. While not everyone in the postmodernist coalition took every batshit antirealist position, they were willing to platform those positions, extend personal solidarity, and provide combative rhetorical cover, ultimately even to pseudoscientific grifters, occultists, new age mystics, faith healers, the Catholic Church, and evangelical creationist. Similarly, many of the science warriors were way too open about associations and collaborations in the name of defeating The Bigger Enemy, a compounding process that turned brains to transphobic and antisemitic goo.

Both sides empowered reactionaries and fascists.

We will be spending generations dealing with the damage, but the mistakes made by both sides were not new or unique to the Science Wars. They were representative of a perpetual tendency within Leftism to hunger for grand coalitions against a single enemy.

Such coalitions necessarily create peripheral zones of sacrifice—both spatial and temporal. Their lossy reduction is measured in a loss of actual lives.

Part of why antifascists were consigned to the fringes of the Left and often ridiculed or attacked for decades before the rise of Trump is that—even putting aside their inherent challenge to the totalitarianism thesis—basic antifascist analyses like Three Way Fight and No Platform are verboten to the

---

232 Meera Nandra, *Prophets Facing Backward* (2004), 221.

Leftist project. No matter how much work antifascists put into keeping other activists safe from fascist street attacks or entryist manipulations, they are always disparaged as wasting time on irrelevancies and undermining unity.

If Leftism is defined as building a pluralistic mass coalition against a single enemy, then the idea of taking threats other than that enemy seriously is unthinkably counter-productive. This is—ironically for their pretensions to represent complexity and dissonance—a place where postmodernism was lapped by the boring old plumbline anarchism that it largely refused to acknowledge or respect.

The antirealists of the Science Wars *did not care* about the oppression that Nanda and I grew up under. They cannot conceive of antirealism as oppressive, they squirm and twist to try to characterize all oppression in terms of hegemonic imposition rather than pluralistic disconnect, but at the end of the day they also simply do not care. We are distant and thus irrelevant concerns. They quickly look past us and walk on. That I was born under the boot of philosophical idealism taken seriously and grew up in the radical left marks me not as an interestingly situated perspective to provide correctives, but as a rounding error.

In contrast to the mass-coalitionist Left, anarchists have long been aware that a very small number of us can—and have repeatedly—changed the world. From this follows the antifascist insight that a small number of out-of-power reactionaries (or statist leftists) can do the same. Our enemies are not singular but manifold.

Radical realists and physicalists are in a very similar position.

At the end of what is broadly recognized as the Science Wars, Latour gave his *mea culpa* and the physicists mostly walked away, declaring victory and considering the matter settled. But their adversaries in the humanities continued redefining and rewriting, building social coalitions and new historical narratives. The postmodernist narratives on the Science Wars have thus been actively constructed, patched, and pushed by a self-perpetuating community in academia whose institutional and economic incentives necessitate the continued creation of such weight. Beyond the continual production of relevant texts, scientists are simply bad at creating the *political coalitions* necessary to retain a basic say in discursive history. Their daily focus is not on mobilizing social allies and their materialist instincts are inherently at odds with liberal democratic pluralism. Relativist sociologists like Fuller were very explicit about heaping praise upon fields like biology, psychology and social sciences with the intention of thus cultivating the broadest possible coalition against the "totalitarian" radicalism of physics.

In this, physicalists could stand to learn from anarchists about what it means to be on an inherently minority footing.

Just as the universalism of physics threatens all other fields and discursive communities, the universalism of anarchism attacks *all* domination, not merely a few specific instances or expressions of it. We can't make exceptions and remain anarchists, and thus we can't promise prospective allies that we'll leave them alone. Our core values offer our enemies nowhere to retreat, and we will always pester and undermine any ally we might make. By definition, every non-anarchist desires *some* form of power, wants to leave *some* power game untouched, and so has reason to unify against us.

Throughout the endless metacommentary on the Science Wars, many noted the difference in expectations and strategies of argument between the two camps. The humanities assume conflict should be a mediative process where both sides massage things into a roughly tolerable peace, where folks learn to live with one another. Whereas the natural scientists always assumed that there will be one correct final conclusion that will for all time stomp the rest into silent oblivion. Social negotiation simply doesn't apply to material facts.

In radical politics, you do frequently just smash your enemy out of existence. If a leninist cult or neonazi crew starts infesting your town, you counter-infiltrate them, map their network, publicly expose them and then physically bust it up and drive them out. This looks like folks going on undercover insertions into their ranks, planting surveillance devices or backdoors, pouring through communications on servers, tearing down their posters, wheatpasting up their information before their neighbors, and slapping the newspapers they're selling out of their hands at protests.

In the Science Wars, the academic postmodernists and relativist sociologists often demanded to know—as if it were unthinkable—if the science warriors thought their departments should be closed up and all of them fired. This was a deliberately unfair rhetorical bait when pretty much every physicist had been emphatic in their support for sociology of science. But, I want to adjust the bounds of the debate by asking the supposedly unthinkable here: *why shouldn't one ideology seek the total annihilation of another ideology?*

What do you think the abolition of chattel slavery amounts to? Do you think it was not built on violence or catalyzed by the extreme actions of a small number of committed radicals? We have built a world in which you can and should be immediately murdered by any decent bystander if they find you attempting to abduct and keep someone as a chattel slave. This is the meat of abolitionism's continuing success. Not any edict preserved on a sheet of paper somewhere that could be torn up tomorrow, but a decentralized moral norm, more deeply and widely embedded. A norm backed by dispersed violent capacity for resistance. The maintenance of this norm is imperfect, but it is broad enough to give pause to would-be slave-owners. One of the greatest advances of liberation and strikes against authoritarianism in human history is underpinned by an imposed universalism that would no doubt be rightfully called decentralized "totalitarian" "absolutism" by some.

Just as we have brutally suppressed the ideology of chattel slavery, we are in the long slog of catalyzing similar norms of resistance against fascists, norms of resistance against domestic abusers, norms of resistance against racists, etc. Such norms are often established, strengthened, and enforced by small minorities who are simply willing to sacrifice more to punish defection from them. Lunch counters and buses were desegregated through the unilateral action of minorities imposing universalistic perspectives, long before the state codified their de facto victories. Such are ceded to, because those seeking power don't care as much and can't organize as well as those willing to sacrifice to stop the domination of others.

If such quintessentially anti-authoritarian approaches are to be described as "totalitarian" then why should anyone care about your definition of "totalitarian"?

> "Anarchism has always been a totalist philosophy."
> Peter Werbe, *Fifth Estate* editor on *Decades in the Struggle*, 2025 [233]

> "Names are indifferent to me; I am not afraid of bugaboos."
> Voltairine de Cleyre & Rosa Slobodinsky,
> *The Individualist and the Communist*, 1891

Why shouldn't we spread similar liberatory norms of resistance against the antirealism of faith healers? When faith healers pressure their dying kids not to go to the emergency room because *"matter is an illusion,"* why shouldn't bystanders beat those parents to death? Or at least otherwise intervene while I choked and gasped on the floor of a supermarket?

And finally, why shouldn't we view academics or leftists arguing for similar forms of antirealism as our ideological enemies? Why shouldn't we see them as actively complicit in murder? Why do we treat Feyerabend defending faith healing or Agamben demanding others not use life-saving medical treatments as some bemusing novelty rather than an ideological force actively murdering people?

If the stakes are *real*, why shouldn't we act like it?

Andrew Ross, an editor of *Social Text*, bemoaned when Sokal made a fool of

---

233 Peter Werbe, "Decades in the Struggle - Interview with Peter Werbe" June 30, 2025, *Nathan Jun*, YouTube, 14:01 https://www.youtube.com/watch?v=yoYVhMd55hI.

him that *"the left eats the left."* The *Social Text* crew endlessly repeated variations of this complaint, and it goes to explain why they gave a pass (and active cover) to quackery and faith healers on the Left, even while such comrades had many of us by the throat. I have no such qualms, and, as an anarchist, no sentimental attraction to *"*the Left,*"* an arbitrarily contingent social coalition and poorly defined aggregate concept with no root if ever there was one.

In the face of some adversaries it is actively counterproductive to be civil or even nonviolent. There comes a point where an unjust peace is worse than a war.

SCIENTISTS AS INSURGENTS

If the academic humanities are rotten with class privilege and reactionary thinking, *rest assured* I have no illusions about actually existing scientific practice today. I treasure many memories like that of my statistical mechanics professor—and head of the department—coming into class gushing about his discovery of Wikipedia, beyond ecstatic with utopian hopes that such a tool would put him out of a job and restore science to the people. But for every People's Science Movement or makeshift commune in the rainforest where anarchist scientists like Grothendieck may retreat, there are thousands of vapid bureaucrats, classroom dictators, and little Eichmanns thoroughly complicit in our hellworld.

I am begging scientists not to identify with the particularities of their employment, but with the radical and transhistorical spirit of their endeavor. Not to be loyal to institutions but to *truth*. To recognize science not as subservient labor or enshrined authority, but as an ideology in the best sense, as a set of driving values.

These ideals are lurking, they have not been entirely killed, but they are being drowned under indoctrinated commitments to civility, democracy, and academia.

Such are not in keeping with the universalism of science. Academia is so often implicitly treated like a transhistorical object but it is nothing of the sort; at best, it comprises a transitive subculture and cluster of organizations that probably won't exist in a few more centuries (maybe even decades). And just as there will come a day when academia is gone, there will come a day when the liberal democratic state is gone.

Universalists should not be wedded to such limited effective domains. The scientific pursuit of truth is a grand project of solidarity across millenia, a collaboration that breaks the zero-sum games of power. What we are building is *a common treasury for all*. And even while power has found some success in containing and channeling science, ultimately science is not in the interests of power.

If pure research is already nearly pressed out of existence, it is plausible, even probable, that within our lives we will see not just mass defundings but proactive state violence against science that extinguishes any place for science in academia or the state. Not just gagging researchers and destroying data, not just teaching lies in schools and expounding them from presidential pulpits, not just enclosing literature in paywalls and blocking access, but active and sustained campaigns of suppression. The arrests and deportations of the second Trump administration, even the bombing of science departments in

Gaza, are not things to be ignored as fringe uncouth violations of civility, they are a glimpse of a war that we are already all a part of and will surely spread.

Science does not provide a retreat of neutrality, it is already a combatant, whether scientists accept this or hide from it.

Contrary to folks like van Fraassen who would reduce science to mere empiricism, science *does* actually reveal content about reality itself. The scientific tradition is not neutral on some cleanly detachable realm of "metaphysics" and it never could be. Both our physicalist model of reality and the theoretical virtues of compression *obviously* prescribe atheism. That's unpopular, but so is radical action on climate change.

Scientists have got to learn to accept being unpopular. They've got to learn to live not in an uneasy subservient marriage with the status quo, but on insurgent footing, fighting for our values in an overwhelmingly hostile world.

The future may well bring centuries of mysticism and social hierarchy, conditions where the remnants of science have to survive and operate in the hidden crevices of the world. The difference between resigned collapse and rebellious flourishing may turn on who thinks of science as a job and who sees it as an orientation and value system. On who thinks of science as an institutional allegiance, and who recognizes it as locked into a war with power.

When indentured workers toiling on algae farms in the arctic pass around subversive literature over illegal encrypted chat apps, what remnant of scientific knowledge will they have? When a radiation-poisoned tribe plants sunflowers to cleanse heavy metals from their soils, what will we contribute to their understanding of an underlying reality?

While a few scientists today smuggle papers on Sci-Hub or lay bombs, more should be done to prepare the insurgent scientists of the future. How we conceptualize ourselves, and how we remember discursive history is an important start.

Ian Hacking has repeatedly registered his disapproval at calling the disagreements over antirealism in the '80s and '90s a "war." He argues that by using the phrase the "Science Wars" we cheapen the horror and seriousness of actual wars. If every social conflict is seen as a war, then when we are on the verge of a real war there is less of a conceptual barrier between cultural or intellectual conflict and outright bloodshed.

I've echoed similar points in critique of anarchists using the phrase "social war" to frame *all* conflict[234] in society as part of an insurgent total war, and, in the process, contributed to a macho fetishization of misanthropic or violent action in sweeping terms. Liberalism's prohibition on violence outside

---

234 Infamously, at one point in the '00s this was taken by a certain writer to include an instance of a worker breaking glass in motel pools so kids would get hurt.

the all-consuming violence of the state is clearly absurd—and patriarchal to its core—but countering such always runs the risk of being interpreted as a blanket endorsement of violence as always superior to persuasion, negotiation, and other pressures.

War has both simplifying associations and pressures. There is always the risk that the banner "war" will be raised only to silence ethical considerations.

Yet, war doesn't have to mean either indiscriminate violence or monolithic armies. Insurgent warfare—like the resistance networks that anarchists built as the nightmare of fascism first rolled across Europe—can recognize and leverage the complexity of the human scale world, surrendering neither to an intolerable peace nor artificial simplification.

Insurgents are often pragmatic, but in a way at least partially inclined to radical realism, searching for deeper dynamics that open possibilities beyond the operating assumptions of a system. An anarchist may find a potential exploit—it could be a software vulnerability on a server, a gap in the search patterns of security guards, a legal loophole in a bank loan scheme, a way to reproduce patented passport ink, etc.—but the risk is steep. If the exploit fails, if the model that suggests its existence is wrong, one can expect to go to prison for at least a decade, perhaps even be executed on the spot. The estimated payoff is determined not merely by the reward or downsides, but by her credence in their model of the potential exploit. Some measure of epistemic caution and diligence is called for, but humility can be a quite substantive *impairment to act*. If such a rebel does not trust in radical analysis of reality, they inflate the likelihood of failure and pass up the opportunity.

Of course there are sharp dangers to overestimating one's credences too—a bomb that ends up slaughtering civilians rather than gendarmerie is worse than inaction—but the postmodern fetishization of humility does not encourage us to studiously strive for *accurate* assessments, it rather bluntly only councils epistemic *underconfidence in general*: it demands that we always doubt our capacity to grasp the real. In such a skewed environment there is no restoring praise for epistemic audacity, nor any microanalysis of cognitive biases and heuristics that could fine-tune us towards a more accurate estimation.

The result, as is so visible with the postmodernists and their children, is self-proclaimed insurgents *who do not act*.

Or who can only attack in the most shallow of instinctive or reactive ways, either passing up the deep modeling necessary to suss out fecund exploits or taking wildly inappropriate gambles that permanently remove them from struggle. Or, worse, collapse them into mere emotive gestures into the void where token violence serves no greater role than social *expression*, devoid of consequence, not even attempting to grip onto the joints of reality or do material damage, closer to a poetry reading or an art school manifesto.

> "Every work of art is an uncommitted crime"
> Theodor Adorno, *Minima Moralia*, 1951

In the end, too much of the Science Wars prioritized performance over action, games of status over any belief in the reality of the stakes. It was a conflict that got people killed, but not the combatants. It littered active ammunition and arms around the planet that continue to shed blood, but in its lasting monuments and mausoleums the only war crimes commemorated are ones of incivility.

It may be that the real "Science Wars" have still yet to take place. I think we should be preparing for them.

We have—as the realist saying goes—a world to win.

Acknowledgements

This book would be entirely impossible without the anarchist movement that I lucked into being born into and whose genius, love, and inspiring sacrifices have buffeted every aspect of my life. There are dozens and dozens of comrades I would proudly die for, who have saved me from prison or worse, and whose examples continuously give me something to live for. I wish I could write you each a note here, but some particularity will have to suffice:

My partner, the notorious graffito criminal, who has personally saved countless lives at immense cost and is the best co-conspirator I could have asked for; our melody is louder than a shout. My bestie, one of the quiet anonymous, principled, brilliant, steadfast tentpoles of the movement who has kept the struggle and the beautiful idea alive under total anonymity. My brother, who is more family than my family and has functionally saved my life enough times to claim any successes I have.

The amazing up-and-coming writer on, among other things, radical scientists and technologists and hackers, Frank Miroslav, for heavy collaboration in research and cleaning up my lazy *"fuck em if they can't google"* sourcing. And who tolerated me rejecting countless additional succulent examples for the sake of concision. Please read his forthcoming books!

The infamous anarchist, mathematician, physicist, neuroscientist, linguist, and whatever else she puts her mind to, Matilde Marcolli, for her early enthusiastic support and foreword.

Julianna Neuhouser and Pete Wolfendale for early feedback. Ky Schevers and RiotLinguist for their insights in the most fringe corners of radical scholarship. Nate Oseroff-Spicer for heavy copyediting and feedback. Scrappy Capy Distro for heavy copyediting. Liam Bright for feedback.

The Internet Archive, without which I would simply have never been able to access hundreds of obscure radical books long out of print, but also Sci-Hub, Anna's Archive, and LibGen. Never forget that the monsters who support intellectual property have names and addresses.

Finally, I'd like to thank my partner's ex, for making the mistake of seething to them about my posts.

BIBLIOGRAPHY

Anarchists Against Democracy: In Their Own Words. Accessed May 2025. https://raddle.me/wiki/anarchists_against_democracy.
Abdurrahman, Fatima. "the physicist who tried to debunk postmodernism." *Dr. Fatima.* Uploaded April 11, 2024. YouTube. https://www.youtube.com/watch?v=ESE-FUaEA7kk.
Abram, David. *The Spell of the Sensuous: Perception and Language in a More-than-Human World.* 1996.
Acker, Kathy. "The gift of disease" *The Guardian.* January 18, 1997. https://outwardfromnothingness.com/the-gift-of-disease-i-el-don-de-la-enfermedad-i/.
Adam, Scott (@ScottAdamSays). "Complexity is always a cover for fraud. In every domain." Twitter, February 7, 2025. https://archive.is/ivPoh.
Adorno, Theodor. *Minima Moralia: Reflections on a Damaged Life.* Translated by E.F.N. Jephcott. 1951.
Ahmed, Sarah. *Differences that Matter: Feminist Theory and Postmodernism.* 1998.
Albert, Michael. "Sokal 1." znetwork. 1996. https://znetwork.org/wp-content/uploads/ScienceWars/sokal_1.htm.
Alvares, Claude. *Science, Development and Violence: The Revolt Against Modernity.* 1992.
Amorós, Miguel. "The Golden Mediocrity." *Libcom.* Translated by Alias Recluse. 2015. https://libcom.org/article/golden-mediocrity-miguel-amoros.
Amster, Randall. "Anarchism as Moral Theory: Praxis, Property, and the Postmodern." *Anarchist Studies* 6, no. 2 (1998): 97–112.
Anderson, P.W. "More is Different." *Science* 177 (1972): 393–396.
Andrews, Mel (@bayesianboy). "I suppose that I diverge from "the methods of a philosopher" in this respect: that I am not interested in deconstructing and debunking the explicitly argued-for stances so much as the pervasive yet unspoken ones." Twitter, 2024. URL lost.
Arendt, Hannah. *Between Past and Future.* 1969.
Arendt, Hannah. *Essays in Understanding.* Edited by Jermone Kohn. 1994.
Arendt, Hannah. *The Origins of Totalitarianism.* 1951.
Arntz, William, Chasse, Betsy and Vicente, Mark, directors. *What the Bleep Do We Know!?* Captured Light, Lord of the Wind. 2004. 109 minutes.
Aronowitz, Stanley. "Conversation with Stanley Aronowitz PhD" Conversations with Harold Hudson Channer. June 26, 1996. Uploaded February 11, 2020. Internet Archive, 46:45. https://archive.org/details/mnn_128_45781995.
Aronowitz, Stanley. "Stanley Aronowitz responds to Norman Levitt." *The Cultural Studies Times* 1, no. 3. (1997) https://web.archive.org/web/19970331132050/http://zelda.thomson.com/routledge/cst/levitt.html.
Aronowitz, Stanley. *Science As Power: Discourse and Ideology in Modern Society.* 1988.
August, Marilyn and Liddle, Ann. "Beyond Structuralism: The Cerisy Experience." *SubStance* 2, no. 5/6, Contemporary French Poetry (Winter, 1972 – Spring, 1973): 227–23.
awkword (@awkword.bsky.social) "Aren't there always multiple perspectives?" Bluesky, January 19, 2025. https://archive.is/OCDKk
awkword (@awkword.bsky.social) "False." Bluesky, January 19, 2025. https://archive.is/A32Bj
Bacus, Florence (@morallawwithin). "If you're a real philosopher then the content of your work should be so deeply central to how you live that any criticism of it feels like a direct personal attack. If you're even capable of having a calm discussion, find another job". Twitter, August 14, 2023. https://archive.is/VcD8V.
Baker, Erik and Oreskes, Naomi. "It's No Game: Post-Truth and the Obligations of Science

Studies." *Social Epistemology Review and Reply Collective* 6, no. 8 (2017): 1–10.
Baker, Erik and Oreskes, Naomi. *Merchants Of Doubt: How a Handful of Scientists Obscured the Truth on Issues from Tobacco Smoke to Global Warming.* 2010.
Bakunin, Mikhail. "Man, Society, and Freedom" in *Bakunin On Anarchy.* Translated and edited by Sam Dolgoff. 1971.
Barnes, Barry and Bloor, David. "Relativism, Rationalism and the Sociology of Knowledge." In *Rationality and Relativism.* Edited by Martin Hollis and Steven Lukes. 1982.
Barnes, Barry, Bloor, David and Henry, John. *Scientific Knowledge: A Sociological Analysis.* 1996.
Barnes, Barry. "Barry Barnes (4 of 4): Rationalisation by experts and their regulators - March 2014 Series." Lecture at ISSTI, March 2014. *The Institute for the Study of Science, Technology and Innovation (ISSTI).* Uploaded September 2, 2014. YouTube, 23:25. https://www.youtube.com/watch?v=yaIM38ux1-Y.
Barnes, Barry. *Realism, Relativism, Finitism.* 1992.
BBC. "Trump aide 'fired over ties to white nationalist event'" August 21, 2018, https://www.bbc.com/news/world-us-canada-45249154.
Beattie, Daren Jeffery. *Martin Heidegger's mathematical dialectic: Uncovering the Structure of Modernity.* Dissertation, Duke University, 2016.
Beiner, Ronald. "The Conservative Revolution of the Twenty-First Century: The Curious Case of Jason Jorjani." In *Contemporary Far-Right Thinkers and the Future of Liberal Democracy.* Edited by A. James McAdams and Alejandro Castrillon. 2022.
Benjamin, Carl (@sargon_of_akkard). "The problem is that reality as we experience it might indeed be radically subjective, and until you can demonstrate that it isn't, then we are in a bind, aren't we?" Twitter, December 7, 2024. https://archive.is/Q5buu.
Berger, Peter L. and Luckmann, Thomas. *The Social Construction of Reality.* 1966.
Berkeley, George. *A Treatise Concerning the Principles of Human Knowledge.* 1710.
Bernal, J.D. *The Social Function of Science.* 1939.
Bernard Iddings, Bell. *Postmodernism and Other Essays.* 1926.
Berube, Michael. "The Science Wars Redux." *Democracy: A Journal Of Ideas* 19 (Winter 2011).
Bey, Hakim. *Immediatism.* 1992.
Bey, Hakim. *Quantum Mechanics & Chaos Theory: Anarchist Meditations on N. Herbert's Quantum Reality: Beyond the New Physics.* Hermetic Library. (Accessed May 2025). https://hermetic.com/bey/quantum.
Bey, Hakim. *T.A.Z.: The Temporary Autonomous Zone, Ontological Anarchy, Poetic Terrorism.* 1991.
Bhaskar, Roy. *A Realist Theory of Science.* 2008.
Bhaskar, Roy. *Reclaiming Reality: A Critical Introduction to Contemporary Philosophy.* 1989.
Bhatt, Talia. ""Sex is Real": The Core of Gender-Conservative Anxiety." In *Trans/Rad/Fem: Essays on Transfeminism.* 2025.
Bhatt, Talia. "The Third Sex" *Trans/Rad/Fem.* September 1, 2024. https://taliabhattwrites.substack.com/p/the-third-sex.
Biggs, Jade. "Grimes says Elon Musk thought she was a "simulation" that he'd created" *Cosmopolitan.* October 17th, 2022. https://www.cosmopolitan.com/uk/entertainment/a41637053/elon-musk-grimes-simulation/.
Bloor, David. "Anti-Lator." *Studies in History and Philosophy of Science Part A* 30, no. 1 (March 1999): 81–112.
Boghossian, Paul. *Fear of Knowledge: Against Relativism and Constructivism.* 2006.
Bollinger, Alex. "Why is Judith Butler trotting out tired excuses to defend a sexual harasser?" *LGBT Nation.* August 14, 2018. https://www.lgbtqnation.com/2018/08/judith-butler-trotting-tired-excuses-defend-sexual-harasser/.
Bookchin, Murray. *Re-enchanting Humanity: A Defense of the Human Spirit Against Anti-Humanism, Misanthropy, Mysticism and Primitivism.* 1995.
Bookchin, Murray. *The Philosophy of Social Ecology: Essays on Dialectical Naturalism.* 1996.
Borges, Jorgre Luis. *The Analytical Language of John Wilkins.* 1942.
Brand, Russell. "The Collective Unconscious, Christ, and the Covenant Russell Brand | EP 444." *Jordan B Peterson.* Uploaded April 30, 2024. YouTube, 30:50. https://youtu.be/

Rl5Z54aYA-E.

Bratton, Benjamin. "Agamben WTF, or How Philosophy Failed the Pandemic" *Verso Books.* July 28th, 2021. https://www.versobooks.com/en-gb/blogs/news/5125-agamben-wtf-or-how-philosophy-failed-the-pandemic.

Bricmont, Jean. *Humanitarian Imperialism: Using Human Rights to Sell War.* Translated by Diana Johnstone. 2006.

Bright, Liam. *"*Empiricism is a Standpoint Epistemology.*"* *The Sooty Empiric.* June 3, 2018. https://sootyempiric.blogspot.com/2018/06/empiricism-is-standpoint-epistemology.html.

Bryant, Levi. "More Remarks on Pluralism: First World Philosophies." *Larval Subjects.* January 24, 2014. https://larvalsubjects.wordpress.com/2014/01/24/more-remarks-on-pluralism-first-world-philosophies/.

Buekens, Filip and Boudry, Maarten. "Psychoanalytic Facts as Unintended Institutional Facts." *Philosophy of the Social Sciences* 24, no. 2 (2011).

Butler, Judith. *Who's Afraid of Gender.* 2024.

Caldwell, Christopher. "Meet the Philosopher Who Is Trying to Explain the Pandemic" *The New York Times.* August 21, 2020. https://www.nytimes.com/2020/08/21/opinion/sunday/giorgio-agamben-philosophy-coronavirus.html.

Carlson, Jedidiah. "Spread This Like Wildfire!" *Science for the People Magazine,* September 26, 2022. https://magazine.scienceforthepeople.org/online/spread-this-like-wildfire.

Chang, Hasok. *Is Water H_2O?: Evidence, Realism and Pluralism.* 2012.

Chase, Alston. *Harvard and the Unabomber: The Education of an American Terrorist.* 2003.

Chopra, Deepak et al. "Why a Mental Universe Is the "Real" Reality." *Chopra Foundation.* Accessed 2024. https://choprafoundation.org/consciousness/why-a-mental-universe-is-the-real-reality/.

Chopra, Deepak. "Everyday Reality is a Human Construct." *Deepak Chopra.* October 24, 2016 https://www.deepakchopra.com/articles/everyday-reality-is-a-human-construct/.

Chopra, Deepak. "The Nature of Reality – Deepak Chopra at MIT." Lecture at MIT, 2018. *The Chopra Well.* Uploaded December 12, 2018. YouTube, 29:45. https://www.youtube.com/watch?v=CHmnPVApfFE.

Clark, Thomas. "Relativism and the Limits of Rationality." *Humanist* 52, no. 1 (1992): 25–32.

Clerk Maxwell, James. "On Faraday's Lines of Force." *Transcriptions of the Cambridge Philosophical Society* 10 (1855): 155–229.

Collard, Andrée and Contrucci, Joyce. *Rape Of The Wild: Man's Violence Against Animals and the Earth.* 1989.

Collier, Andrew. *Critical Realism: An Introduction to Roy Bhaskar's Philosophy.* 1994.

Collins, Harry. "Stages in the Empirical Program of Relativism." *Social Studies of Science* 11, no. 1 (1981): 3-10.

Collins, Harry. *Changing Order: Replication and Induction in Scientific Practice.* 1992.

Copernicus, Nicolaus. *On the Revolutions of the Celestial Spheres.* Translated Edward Rosen. 1978.

CrimethInc. *From Democracy To Freedom: The difference between government and self-determination.* 2017.

Cromer, Alan. *Connected Knowledge: Science, Philosophy and Education.* 1997.

Cusset, Françios. *French Theory: How Foucault, Derrida, Deleuze, & Co. transformed the intellectual life of the United States.* Translated by Jeff Fort, Josphine Berganza and Marlon Jones. 2008.

David Mermin, N. *Boojums All the Way Through: Communicating Science in a Prosaic Age.* 1990.

Dawid, Richard. *String Theory and the Scientific Method.* 2013.

de Cleyre, Voltairine and Slobodinsky, Rosa. "The Individualist and the Communist." *The Twentieth Century* 15. June 18, (1891).

de Lauretis, Teresa. "The violence of rhetoric: Considerations on representation and gender" *Semiotica* 54 no. 1/2 (1985): 11–31.

DeDeo, Simon. *Mutual Explainability, or, A Comedy in Computerland.* April 19, 2020. http://fqxi.data.s3.amazonaws.com/data/essay-contest-files/DeDeo_Comedy_in_Computerlan.pdf.

DeLanda, Manuel. "Manuel Delanda, Deleuze and the Use of the Genetic Algorithm in Architecture." Columbia Art and Technology Lectures. April 9, 2004. *Columbia University*. Uploaded April 29, 2009. YouTube, 4:20. https://www.youtube.com/watch?v=50-d_J0hKz0.

Derrida, Jacques. "Letter from Jacques Derrida to Ralph J. Cicerone, then Chancellor of UCI." July 25, 2004. https://web.archive.org/web/20190131152027/http://www.jacques-derrida.org/Cicerone.html.

Derrida, Jacques. "Sokal et Bricmont ne sont pas sérieux." *Le Monde*. November 20, 1997. https://web.archive.org/web/20090216071032/http://peccatte.karefil.com/SBPresse/LeMonde201197Derrida.html.

Derrida, Jacques. *Paper Machines*. Translated by Rachel Bowlby. 2005.

Derrida, Jacques. *Resistances of Psychoanalysis*. Translated by Peggy Kamuf, Pascale-Anne Brault, and Michael Naas. 1998.

Derrida, Jacques. *Writing And Difference*. 1967.

Dosse, Francios. *History of Structuralism: Volume 1 The Rising Sign, 1945–1966*. Translated by Deborah Glassman. 1997.

Drake, James. "The Academic Brand of Aphasia: Where Postmodernism and the Science Wars Came From." *Knowledge, Technology and Policy* 15 (2002): 13–187.

Dugin, Alexander. "Unregistered 193: Alexander Dugin (VIDEO)." *Unregistered Podcast*. Uploaded January 7, 2022. YouTube, 7:43. https://www.youtube.com/watch?v=X2RxutaSGQU.

Duhem, Pierre. *The Aim and Structure of Physical Theory*. Translated by Philip P. Wiener. 1954.

Dupré, John. *The Disorder of Things: Metaphysical Foundations of the Disunity of Science*. 1995.

E. Babich, Babette. "Physics vs. Social Text: Anatomy of a Hoax." *Telos* 1996 no. 107 (1996): 45–61.

Eagleton, Terry. *After Theory*. 2003.

Editorial. "Professor Sokal's Transgression" *The New Criterion*. June, 1996. https://newcriterion.com/article/professor-sokals-transgression/.

Einstein, Albert and Infeld, Leopold. *The Evolution of Physics: The Growth of Ideas from Early Concepts to Relativity and Quanta*. 1938.

Ellul, Jacques. *Anarchy and Christianity*. 1991.

Epstein, Barbara. "Postmodernism and the Left." *New Politics* 6, no. 2 (1997).

Eumaios. "Objective Fictions & Subjective Realities: The Need for a Nationalist Postmodernism." *Counter Currents*. May 5, 2020. https://counter-currents.com/2020/05/objective-fictions-subjective-realities-the-need-for-a-nationalist-postmodernism/.

Feyerabend, Paul. *Against Method: Outline of an Anarchistic Theory of Knowledge*. 1st ed. 1975.

Feyerabend, Paul. *Against Method: Outline of an Anarchistic Theory of Knowledge*. 3rd ed. 1993.

Feyerabend, Paul. *Farewell to Reason*. 1988.

Feyerabend, Paul. *Killing Time: The Autobiography of Paul Feyerabend*. 1994.

Feyerabend, Paul. *Science in a Free Society*. 1978.

Feyerabend, Paul. *The Tyranny Of Science*. 1996.

Feyerabend, Paul. "Atoms and Consciousness." *Common Knowledge* 1 (1992): 28–32.

Fink, Bruce. *The Lacanian Subject: Between Language and Jouissance*. 1995.

Firestone, Shulamith. *The Dialectic of Sex: The Case for Feminist Revolution*. 1970.

Fish, Stanley. "Consequences." *Critical Inquiry* 11 no. 3 (1985): 433–458.

Fish, Stanley. "Professor Sokal's bad joke" *New York Times*. May 21, 1996. https://www.nytimes.com/1996/05/21/opinion/professor-sokal-s-bad-joke.html.

Fisher, Mark. "Terminator vs Avatar." In *#ACCELERATE: The Accelerationist Reader*, edited by Robin MacKay and Armen Avanessian. 2014.

Fitzpatrick, Bellamy. *Corrosive Consciousness*. 2017.

Flores Magon, Ricardo. "Preaching Peace is a Crime" *Regeneration*. Machine translated. September 17, 1910. https://www.antorcha.net/biblioteca_virtual/politica/ap1910/5.html.

Foucault, Michel. "Truth and Juridical Forms." In *Power: Essential Works of Foucault, 1954–1984: Volume Three*. Translated by Robert Hurley. Edited by James D. Faubion. 2000.

Foucault, Michel. *Madness And Civilization: A History of Insanity in the Age of Reason*. 1961.
Foucault, Michel. *The Order Of Things: An Archaeology of the Human Sciences*. 1966.
Fraenkel, Abraham A. "How German Mathematicians Dealt With the Rise of Nazism" *Tablet*, February 8th, 2017. https://www.tabletmag.com/sections/arts-letters/articles/hitlers-math.
Franklin, Sarah. "Making Transparencies: Seeing through the Science Wars." Social Text 46/47, Science Wars (Spring - Summer, 1996): 141–155.
Frawley, David. *Awaken Bharat: A Call for India's Rebirth*. 1998.
Fromm, Harold. "My Science Wars." *The Hudson Review* 49, no. 4 (Winter, 1997): 599–609.
Fuentes, Agustín. "Here's Why Human Sex is not Binary" *Scientific American*. May 1, 2023. https://www.scientificamerican.com/article/heres-why-human-sex-is-not-binary/.
Fuller, Steve. "Is STS All Talk and No Walk?" *East Review* 36 no. 1 (2017): 21–23.
Fuller, Steve. *Science: Concepts in Social Sciences*. 1997.
G. Wilson, Kenneth and K. Barsky, Constance. "Beyond Social Construction". In *The One Culture?* Edited by Jay A. Labinger and Harry Collins. 2001.
Galilei, Galileo. *Dialogue Concerning the Two Chief World Systems*. Translated by Stillman Drake. 1953.
Gemie, Sharif. "Habermas and Anarchism." *Bulletin of Anarchist Research (UK)* 24 (Summer, 1991): 18–20.
Gergen, Kenneth J. "Social Psychology as History." *Journal of Personality and Social Psychology* 26, no. 2 (1973): 309–320.
Gergen, Kenneth. "Feminist critique of science and the challenge of social epistemology." In Feminist thought and the structure of knowledge. Edited by M. Gergen. 1998.
Gillis, William. "A Giant Red Flag Folded Into a Book." *Human Iterations*. December 6, 2022. https://humaniterations.net/2022/12/6/giant-red-flag.
Gillis, William. "Anti-Engels (or Anti-Anti-Duhring Aktion)." *Human Iterations*. November 5, 2021. https://humaniterations.net/2021/11/5/anti-engels.
Gillis, William. "Antifa Activists as the Truest Defenders of Free Speech." *Human Iterations*. November 19, 2017. https://c4ss.org/content/50151.
Gillis, William. "From Stirner to Mussolini: Review: The Anarchist-Individualist Origins of Italian Fascism." *Human Iterations*. March 28, 2022. https://humaniterations.net/2022/03/28/stirner-mussolini.
Gillis, William. "Responding to Fascist Organizing." *Center for a Stateless Society*. January 25, 2017. https://c4ss.org/content/47734.
Gillis, William. "Science as Radicalism." *Human Iterations*. August 18, 2015. https://humaniterations.net/2015/08/18/science-as-radicalism/.
Gillis, William. "The Abolition Of Rulership Or The Rule Of All Over All?" *Human Iterations*. June 12, 2017. https://humaniterations.net/2017/06/12/the-abolition-of-rulership-or-the-rule-of-all-over-all/.
Gillis, William. "What's In A Slogan? 'Kill Your Local Rapist' and Militant Anarcha-feminism." *Human Iterations*. June 25, 2024. https://humaniterations.net/2024/06/25/kylr/.
Goldman, Alvin. *Knowledge in a Social World*. 1999.
Gottfried, Kurt. "Was Sokal's Hoax Justified?" *Physics Today* 50 no.1 (1997): 61–62.
Graeber, David. *Anarchy - In a Manner of Speaking: Conversations with Mehdi Belhaj Kacem, Nika Dubrovsky, and Assia Turquier-Zauberman*. 2020.
Graeber, David. "Dead zones of the imagination: on violence, bureaucracy, and interpretive labor. The 2006 Malinowski Memorial Lecture" *HAU: Journal of ethnographic theory* 2 no. 2 (2012): 105–128.
Graeber, David. *Fragments of an Anarchist Anthropology*. 2004.
Greenhouse, Linda. "Christian Scientists Rebuffed in Ruling By Supreme Court" *The New York Times*. January 23rd, 1996. https://www.nytimes.com/1996/01/23/us/supreme-court-roundup-christian-scientists-rebuffed-in-ruling-by-supreme-court.html.
Greer, Joseph Christian. "Occult Origins: Hakim Bey's Ontological Post-Anarchism." *Anarchist Developments in Cultural Studies* 2 (2014): 168–187.
Gross, Paul R. and Levitt, Norman. *Higher Superstition: The Academic Left and Its Quarrels*

with Science. 1994.
H. Barlow Ph.D., David, G. Abel M.D., Gene and B. Blanchard, Edward. "Gender identity change in a transsexual: An exorcism." *Archives of Sexual Behavior* 6 (1977): 387–395.
Haack, Susan. *Defending Science Within Reason: Between Scientisim and Cynicism*. 2007.
Habermas, Jürgen. "Modernity versus Postmodernity." *New German Critique* 22, Special Issue on Modernism (Winter, 1981): 3–14.
Hahn, Julia. "Consciousness and Society." spi conference i. August 19, 2013. *SPI — The Society for Psychoanalytic Inquiry*. Uploaded November 25, 2013. YouTube, 48:23 https://youtu.be/39FMKMdoM18.
Hahn, Julia. "Slovenian Marxist Philosopher Slavoj Žižek Says He'd Vote Trump: Hillary Clinton 'Is the Real Danger'" *Breitbart*. November 4, 2016. https://www.breitbart.com/politics/2016/11/04/slavoj-zizek-vote-trump-hillary-real-danger/.
Hancock, Graham. "Writing about Outrageous Hypotheses and Extraordinary Possibilities: A View from The Trenches." *Graham Hancock*. January 19, 2002. https://grahamhancock.com/outrageous-hypotheses-hancock/.
Harman, Graham. "The Importance Of Bruno Latour For Philosophy." *Cultural Studies Review* 13 no. 1 (2007):31–49.
Harper's Magazine. *A Letter on Justice and Open Debate*. July 7, 2020. https://harpers.org/a-letter-on-justice-and-open-debate/.
Heisenberg, Werner. *Physics and Philosophy: The Revolution in Modern Science*. 1958.
Hesse, Mary. *Revolutions and Reconstructions in the Philosophy of Science*. 1980.
Hicks, Stephen. *Explaining Postmodernism: Skepticism and Socialism from Rousseau to Foucault*. 2004.
Hindess, Barry and Hirst, Paul. *Mode of Production and Social Formation: An Auto-Critique of Pre-Capitalist Modes of Production*. 1977.
Hipwell, Kim. "The Matter Of "Material Girls": Conspiracy Theory In Anti-Trans Philosophy." *Medium*. March 12, 2023. https://kim-hipwell.medium.com/the-matter-of-material-girls-71bf7a488fbe.
Hobsbawm, Eric. *How to Change the World: Reflections on Marx and Marxism*. 2012.
Horkheimer, Max and Adorno, Theodore. *Dialectic of Enlightenment*. Edited by Gunzelin Schmid Noerr. Translated by Edmund Jephcott. 2002.
Huemer, Michael. *Progressive Myths*. 2024.
J.F. Day, Richard. *Gramsci is Dead: Anarchist Currents in the Newest Social Movements*. 2005.
Lindsay, James. (@ConceptualJames). "If this ideology isn't stopped, one way or another, there will be. Ideally, we just start saying no and push them out with legal means. Their funders might liquidate or mass incarcerate them when they seize power too, which is way less good but avoids it." Twitter, June 10, 2021. https://web.archive.org/web/20210628000113/https://twitter.com/ConceptualJames/status/1402749362220974094.
Lindsay, James (@ConceptualJames). "If you didn't know, "the plan" involves reducing the world population to under 2B, perhaps by the end of the decade. That's about 3/4 people who are excess population." Twitter, January 13, 2021. https://web.archive.org/web/20210324145446/https://twitter.com/ConceptualJames/status/1349370680857518082.
Lindsay, James (@ConceptualJames). "Carrying the flag of a hostile enemy in the military. Shame." Twitter, June 17, 2020. https://archive.is/8geaj.
James, Joy. "The Architects of Abolitionism." The Architects of Abolitionism: George Jackson, Angela Davis, and the Deradicalization of Prison Struggles lecture. April 8, 2019. *Brown University*. Uploaded May 7, 2019. YouTube, 39:35 https://www.youtube.com/watch?v=z9rvRsWKDx0.
James, William. "The Will to Believe: Address to the Philosophical Clubs of Yale and Brown Universities" *the New World*. June, 1896. https://arquivo.pt/wayback/20090714151749/http://falcon.jmu.edu/~omearawm/ph101willtobelieve.html.
Jasanoff, Sheila. "Beyond Epistemology: Relativism and Engagement in the Politics of Science." *Social Studies of Science* 26, no. 2 (1996): 393–418.
Jean Bricmont Personality Notice. *Conspiracy Watch*. Accessed February 2025. https://www.

conspiracywatch.info/notice/jean-bricmont.
Jern, Alan, Chang, Kai-min K., and Kemp, Charles. "Belief polarization is not always irrational." *Psychological Review* 121 no. 2 (2014): 206–224.
Johnson, Sonia. *Going Out Of Our Minds: The Metaphysics of Liberation.* 1987.
Johnson, Sonia. *The SisterWitch Conspiracy.* 2010.
k-dog. "Fifth Column Fascism: fascism within the anti-war movement." *ARA Research Bulletin* 2 (Fall, 2001).
Kaiser, David. "History: Shut up and calculate!" *Nature* 505, (2014): 153–155.
Kaiser, David. "Nuclear Democracy: Political Engagement, Pedagogical Reform, and Particle Physics in Postwar America." *Isis* 93, no. 2 (2002): 229–268.
Kaiser, David. "The Atomic Secret in Red Hands? American Suspicions of Theoretical Physicists During the Early Cold War." *Representations* 90, no. 1, (2005): 28–60.
Kaiser, David. *How the Hippies Saved Physics: Science, Counterculture, and the Quantum Revival.* 2012.
Kant, Immanuel. *Critique of Pure Reason.* Translated by J. M. D. Meiklejohn. 1787.
Kant, Immanuel. *Kant's Inaugural Dissertation of 1770.* Translated by William J. Eckoff. 1894.
Kircher, Philip. *Science, Truth, Democracy.* 2001.
Kissinger, Henry. *American Foreign Policy: Three Essays.* 1969.
Knight, Michael Muhammad. *William S. Burroughs vs. the Qur'an.* 2012.
Koch, Andrew M. "Poststructuralism and the Epistemological Basis of Anarchism." *The Philosophy of the Social Sciences* 23, no. 3 (September 1993): 327–35.
Kotsko, Adam. "What Happened to Giorgio Agamben?" *Slate.* February 20, 2022. https://slate.com/human-interest/2022/02/giorgio-agamben-covid-holocaust-comparison-right-wing-protest.html.
Kucsh, Martin. *Relativism in the Philosophy of Science.* 2020.
Kuhn, Thomas. *Logic Of Discovery Or Psychology Of Research?* Edited by Imre Lakatos and Alan Musgrave. 1970.
Kuhn, Thomas. *The Structure Of Scientific Revolutions.* 1970.
Kukla, Andre. "Does Every Theory Have Empirically Equivalent Rivals?" *Erkenntnis* 44 (1996): 137–166.
Kusch, Martin. *Relativism in the Philosophy Of Science.* 2020.
L. Allen, John, Jr. "Ratzinger's 1990 remarks on Galileo" *National Catholic Reporter.* January 14, 2008 https://www.ncronline.org/news/ratzingers-1990-remarks-galileo.
Lacan, Jacques. *Desire and Its Interpretation: The Seminars of Jacques Lacan.* Translated Cormac Gallagh. 1958.
Lacan, Jacques. *Écrits.* Translated by Bruce Fink. 2007.
Laclau, Ernesto. "Universalism, Particularism and the Question of Identity." *October* 61, The Identity in Question (Summer, 1992): 83–90.
Lakatos, Imre. "History Of Science and its Rational Reconstructions." In *PSA 1970: In Memory of Rudolf Carnap Proceedings of the 1970 Biennial Meeting Philosophy of Science Association.* Edited by Roger C. Buck, Robert S. Cohen. 1971.
Lamborn Wilson, Peter and Pourjavady, Nasrollah. *Kings of Love.* 1978.
Land, Nick. "Against Universalism." In *Xenosystem Fragments.* Edited by Apostate. 2016.
Landstreicher, Wolfi. "A Balanced Account of the World: A Critical Look at the Scientific World View." *Killing King Abacus* 2 (2001): 44–49.
Landstreicher, Wolfi. *Rants, Essays and Polemics of Feral Faun.* 1987.
Langer, Elinor. *A Hundred Little Hitlers: The Death of a Black Man, the Trial of a White Racist, and the Rise of the Neo-Nazi Movement in America.* 2004.
Larson, Cynthia Sue. *Reality Shifts: When Consciousness Changes the Physical World.* 1991.
Latour, Bruno. "For David Bloor… and Beyond: A Reply to David Bloor's 'Anti-Latour.'" *Studies in history and philosophy of science* 30, no. 1 (1999) 113–129.
Latour, Bruno. "On the Partial Existence of Existing and Non-existing Objects." In *Biographies of Scientific Objects.* Edited by Lorraine Daston. 1996.
Latour, Bruno. "Why has Critique Run Out of Steam?" *Critical Inquiry* 30, no. 2 (Winter 2004): 225–248.

Latour, Bruno. *Science in Action.* 1987.
Latour, Bruno. *The Pasteurization of France.* Translated by Alan Sheridan and John Law. 1998.
Laudan, Larry. *Science And Relativism.* 1990.
Lederman, Leon M. and Hill, Christopher T. *Symmetry and the Beautiful Universe.* 2004.
Lefort, Claude. "Politics and Human Rights." In *The Political Forms of Modern Society: Bureaucracy, Democracy, Totalitarianism.* Edited by John B. Thompson. 1986.
Leigh, James. "Free Nietzsche." *Semiotext(e)* III, no. 1 (1997): 4–6.
Levitt, Norman. "Why Professors Believe Weird Things: Sex, Race and the Trials of the New Left." *eSkeptic.* January 14, 2008. https://archive.skeptic.com/archive/eskeptic/14-01-08/#feature.
Lincoln, Yvonna and Guba, Egon. *Naturalistic Inquiry.* 1985.
Lindsay, James. "Applied Postmodernism: How "Idea Laundering" is Crippling American Universities." Aspen Jewish Community Center. July 30, 2019. *GrassRoots Community Network.* Uploaded August 9, 2019. YouTube, 44:45. https://www.youtube.com/watch?v=XeXfV0tAxtE.
Lipstadt, Deborah. *Denying the Holocaust: The Growing Assault on Truth and Memory.* 1993.
Luu, Thomas and Meissner, Ulf-G. "On the Topic of Emergence from an Effective Field Theory Perspective." In *Top-Down Causation and Emergence.* Edited by Jan Voosholz, Markus Gabriel. 2021.
Luz (@likelyspam.bsky.social) "What is false? I work there lmao." Bluesky, January 19, 2025. https://archive.is/d76wq.
Luz (@likelyspam.bsky.social) "They're calling the Black and brown workers behind this colonizers. And getting white men like you to spread this message. Disgusting." Bluesky, January 19, 2025. https://archive.is/Vt3Xc
Lyotard, Jean-François. *Driftworks.* Edited by Roger McKeon. 1984.
Lyotard, Jean-François. *Dérive à partir de Marx et Freud.* 1973.
Lyotard, Jean-François. *Libidinal Economy.* 1974.
Lyotard, Jean-François. *The Postmodern Condition: A Report on Knowledge.* Translated by Geoff Bennington and Brian Massumi. 1984.
Machamer, Peter K. "Feyerabend and Galileo: The interaction of theories, and the reinterpretation of experience." *Studies in History and Philosophy of Science Part A* 4 no. 1 (1973):1–46.
Macklin, Graham. "Greg Johnson and Counter- Currents." In *Radical Thinkers of the New Right: Behind the New Threat to Liberal Democracy.* Edited by Mark Sedgwick. 2019.
Mae Bettcher, Talia. ""When Tables Speak": On the Existence of Trans Philosophy" *Daily Nous.* May 30, 2018. https://dailynous.com/2018/05/30/tables-speak-existence-trans-philosophy-guest-talia-mae-bettcher/.
Mangan, Katherine. "New Disclosures About an NYU Professor Reignite a War Over Gender and Harassment" *The Chronicle of Higher Education.* August 15, 2018. https://www.chronicle.com/article/new-disclosures-about-an-nyu-professor-reignite-a-war-over-gender-and-harassment/.
Marantz, Andrew. "Trolls for Trump" *The New Yorker.* October 24, 2016. https://www.newyorker.com/magazine/2016/10/31/trolls-for-trump.
Marin, Emmanuel. *Summary of Articles from Le Monde.* Sokal & Social Text. February 25, 1997. https://web.archive.org/web/20050214122551/http://www.drizzle.com/~jwalsh/sokal/articles/emarin.html.
Mas, John. *Communication and Control.* Ph.D., University of Hawaii at Manoa. 2014.
Maxwell Clerk, James. "On the Dynamical Evidence of the Molecular Constitution of Bodies." *Nature* 11 (1875): 357–359.
May, Todd. *The Moral Theory of Poststructuralism.* 2004.
May, Todd. *The Political Philosophy of Poststructuralist Anarchism.* 1994.
McConnell, Craig and H. March, Robert. "Bringing Reason and Context to the Science Wars" *Physics Today* 54 no. 5 (2001):57–58.
McKeon, Roger. "Gaiety, A Difficult Science." *Semiotext(e)* III, no. 1 (1997): 8–11.
Meier, Richard L. "The Origins of the Scientific Species." *Bulletin of the Atomic Scientists* 7,

no. 6 (1951): 169–173.
Meillassoux, Quentin. *After Finitude: An Essay on the Necessity of Contingency.* Translated by Ray Brassier. 2009.
Merchant, Carolyn. *The Death of Nature: Women, Ecology and the Scientific Revolution.* 1980.
Merleau-Ponty, Maurice. *Phenomenology of Perception.* Translated by Colin Smith. 1945.
Michel, Foucault. *Power/Knowledge: Selected Interviews and Other Writings, 1972-1977*, ed. Colin Gordon, trans. Colin Gordon, et al. 1988.
Mies, Maria and Shiva, Vandana. *Ecofeminism.* 1993.
Morrey, Douglas. "Manifeste conspirationniste, Parti imaginaire, Comité invisible: a genealogy of radical critique in twenty-first-century France." *Modern & Contemporary France* 33, no. 2 (2025): 209–224.
Morris, Erol. *The Ashtray: (Or the Man Who Denied Reality).* 2018.
Mukhyananda, Swami. *Vedanta in the context of modern science: A comparative study.* 1997.
Munzi, Ulderico. "'Francesi, intellettuali impostori': Americani all'attacco di Parigi'" *Corriere della Sera.* September 26, 1997. http://www.symbolic.parma.it/bertolin/ms11.htm.
Namaste, Viviane. "Undoing Theory: The "Transgender Question" and the Epistemic Violence of Anglo-American Feminist Theory." *Hypatia* 24, no. 3, Transgender Studies and Feminism: Theory, Politics, and Gendered Realities (2009): 11–32.
Nanda, Meera. *Prophets Facing Backward: Postmodern Critiques of Science and Hindu Nationalism in India.* 2004.
Nandy, Ashis and Visvanathan, Shiv. "Modern Medicine and its Non-Modern Critics: A Study in Discourse." In *Dominating Knowledge: Development, Culture, and Resistance.* Edited by Frédérique Apffel Marglin and Stephen A. Marglin. 1990.
Nelkin, Dorothy. "The Science Wars: Responses to a Marriage Failed." Social Text 46/47, Science Wars (Spring - Summer, 1996): 93–100.
Newman, Saul. *The Politics of Postanarchism.* 2010.
Newport, Frank and Dugan, Andrew. "College-Educated Republicans Most Skeptical of Global Warming" *Gallup.* March 26, 2015. https://news.gallup.com/poll/182159/college-educated-republicans-skeptical-global-warming.aspx.
Nkrumah Kwame. *Neo-colonialism: The Last Stage of Imperialism.* 1965.
O'Meara, Michael. *New Culture, New Right: Anti-Liberalism in Postmodern Europe.* 2013.
OEIS Foundation Inc. The Dounda Sequence. Entry A005940 in The On-Line Encyclopedia of Integer Sequence. Accessed 2025. https://oeis.org/A005940.
Ōshima, Nagisa, director. *Death by Hanging.* Sozosha. Art Theatre Guild, 1968. 118 minutes.
Ove Hansson, Sven. "Social constructionism and climate science denial." *General Philosophy of Science* 10, no. 3 (2020): 1–27.
Parisi, Giorgio. "Nobel Lecture: Multiple equilibria." Preprint. Submitted April 2, 2023. https://arxiv.org/abs/2304.00580.
Peter Lamborn Wilson. "Boundary Violations." In *Technoscience and Cyberculture.* Edited by Stanley Aronowitz. 1996.
Peterson, Christa. "Kathleen Stock, OBE." *praile.* January 21, 2021. https://www.praile.com/post/kathleen-stock-obe.
Peterson, Jordan. "Jordan Peterson Exposes the Postmodern Agenda (Part 1 of 7)." *Epoch Times.* Uploaded July 4, 2017. YouTube, 1:00. https://www.youtube.com/watch?v=P-kNzYttjSHE.
Peterson, Jordan. "Sam Harris vs Jordan Peterson | God, Atheism, The Bible, Jesus – Part 4 - Presented by Pangburn" O2 Arena in London, England. July 16, 2018. *Pangburn.* Uploaded September 14, 2018. YouTube, 1:01:00. https://www.youtube.com/watch?v=aALsFhZKg-Q.
Poor, Jerod. "Nihilism." *Popular Reality* 7 (1985): 3. https://www.jchristiangreer.com/s/PopRealNo7.pdf.
Postrel, Steven and Feser, Edward. "Reality Principles: An Interview with John R. Searle" *Reason Magazine.* February 1, 2000. https://reason.com/2000/02/01/reality-principles-an-intervie/.
Prescod-Weinstein, Chanda. *The Disordered Cosmos: A Journey into Dark Matter, Spacetime,*

and Dreams Deferred. 2021.

Pseudo-chrysostom. Comment on "The Disastrous Effects Of Females In Power." *Jim's Blog*. June 5, 2018. https://blog.reaction.la/economics/the-disastrous-effects-of-females-in-power/#comment-1782721.

Putnam, Hillary. *Realism with a Human Face*. 1990.

Radder, Hans. "Empiricism must, but cannot, presuppose real causation." *Journal for General Philosophy of Science* 52 (2021): 597–608.

Rauschning, Herman. *Hitler Speaks: A Series Of Political Conversations With Adolf Hitler On His Real Aims*. 1940.

Ravetz, Jerry. *Scientific Knowledge and its Social Problems*. 1971.

Raynor, Alexander. "Meeting with the Father of Neoreaction." *European New Right Revue*. February 12, 2025. https://nouvelledroite.substack.com/p/meeting-with-the-father-of-neoreaction.

Reno, R.R. "A Responsible Bishop" *First Things*. April 17, 2020. https://firstthings.com/a-responsible-bishop/.

Ricos, los Rob. "Law and Disorder @ PSU - Rob los Ricos Thaxton." Law and Disorder Conference at Portland State University. 2009. Uploaded November 18, 2013. Internet Archive, 11:00 https://archive.org/details/bliptv-20131014-004423-Bmediacollective-LADRobLosRicosH26424fps528.

Rivenburg, Roy. "Were sex and punishment behind feud for archives?" *Los Angeles Times*. February 25, 2007. https://www.latimes.com/archives/la-xpm-2007-feb-25-me-derrida25-story.html.

Robbins, Bruce. "Anatomy of a Hoax." *Tikkun* September/October (1996): 58-59.

Roderick, Rick. "Nietzsche on Truth and Lie [full length]." Nietzsche and the Postmodern Condition series, 1991. *The Partially Examined Life*. Uploaded January 28, 2012. YouTube, 7:22. https://www.youtube.com/watch?v=KyOP2D5H0nw.

Rorty Richard. *Philosophy and the Mirror of Nature*. 1979.

Rorty, Richard. "Pragmatism as Antiauthoritarianism" Part 1. Richard Rorty's 1996 Girona Lectures, with discussions. Richard Rorty" Ferrater Mora lectures, University of Girona, 1996. *Bob Bradom*. Uploaded November 21, 2021. YouTube, 4:18, 26:0 and 1:51:32 . https://www.youtube.com/watch?v=-k-IUoEAHpg.

Rorty, Richard. *Philosophy and Social Hope*. 1999.

Rorty, Richard. "The Contingency of Language." *London Review of Books* 8 no. 7 (17 April 1986).

Rorty, Richard. *Truth and Progress: Volume 3: Philosophical Papers Volume 3*. 1998.

Rorty, Richard. *Pragmatism as Anti-Authoritarianism*. 2021.

Rosen, Jay. "Swallow Hard: What Social Text Should Have Done." *Tikkun* September/October (1996): 59–61.

Ross, Andrew. "Introduction." *Social Text* 46/47 Science Wars (spring/summer 1996): 1–15.

Ross, Andrew and Robins Bruce. "Mystery Science Theatre." *Lingua Franca* (July/August 1996): 54–57.

Roush, Sherri. "Optimism about the Pessimistic Induction." In *New Waves in Philosophy of Science*. Edited by PD Magnus and Jacob Buschby (2010).

Rousselle, Duane. "Post-Anarchism and Psychoanalysis." *Cyber Dandy*. Uploaded July 28, 2023. YouTube, 29:10. https://www.youtube.com/watch?v=S0pz5SBHIQ8.

Salmon, Peter. "After Jacques Derrida, What's Next For French Philosophy?" *Aeon Essays*. May 6, 2022. https://aeon.co/essays/after-jacques-derrida-whats-next-for-french-philosophy.

Sardar, Ziauddin. *Postmodern Encounters: Thomas Kuhn and the Science War*. 2002.

Schindler, Samuel. *Theoretical Virtues in Science: Uncovering Reality through Theory*. 2018.

Scott, James C. *Seeing Like a State: How Certain Schemes to Improve the Human Condition Have Failed*. 1996.

Serano, Julia. "What is a Woman? (a response)." *Medium*. September 28, 2023. https://juliaserano.medium.com/what-is-a-woman-a-response-8f91aaf3a971.

Shackel, Nicholas. "The Vacuity of Postmodern Methodology." *Metaphilosophy* 36 (April 2005): 295–320.

Shapin, Steven and Schaffer, Simon. *Leviathan and the Air-Pump: Hobbes, Boyle, and the Experimental Life*. 1985.

Sharma, Nandita. *Home Rule: National Sovereignty and the Separation of Natives and Migrants*. 2020.

Shermer, Michael. "Farewell to Norman Jay Levitt (1943–2009)" *eSkeptic*. October 27, 2009. https://archive.skeptic.com/archive/eskeptic/09-10-26/.

Shiva, Vandana. "Reductionist science as epistemological violence." In *Science, hegemony and violence: a requiem for modernity*. Edited by Ashis Nandy. 1988.

Shlain, Leonard. *The Alphabet and the Goddess: The Conflict Between Word and Image*. 1998.

Simmons, Jonathan. "Indigenous atheism and the spiritual stereotype" *OnlySky Media*. June 29, 2022. https://web.archive.org/web/20220708172643/https://onlysky.media/jsimmons/indigenous-atheism-and-the-spiritual-stereotype/.

Smith, Blake. "Bronze Age Pervert's Dissertation on Leo Strauss" *Tablet*. February 15, 2023. https://www.tabletmag.com/sections/arts-letters/articles/bronze-age-pervert-dissertation-leo-strauss.

Smith, Blake. "The Inner Life of Gender" *Tablet*. February 15, 2023. https://www.tabletmag.com/sections/arts-letters/articles/gender-neutral.

Smith, Blake. "The unwoke Foucault" *Washington Examiner*. March 5, 2021. https://www.washingtonexaminer.com/magazine-life-arts/2227707/the-unwoke-foucault/.

Snow, C.P. *The Two Cultures*. 1998.

Sokal, Alan and Bricmont, Jean. *Fashionable Nonsense: Postmodern Intellectuals' Abuse of Science*. 2nd edition. 1999.

Sokal, Alan. "Preface to the French translation of Helen Pluckrose and James Lindsay." In Helen Pluckrose and James Lindsay *Cynical Theories: How Activist Scholarship Made Everything About Race, Gender, and Identity*. 2021. https://physics.nyu.edu/faculty/sokal/preface_to_cynical_theories_ENGLISH.pdf.

Sokal, Alan. "Submission to the Government Consultation on "Banning Conversion Therapy." December 5, 2021. https://physics.nyu.edu/faculty/sokal/conversion_therapy_submission.pdf.

Sokal, Alan. "The Bad Faith Use Of Words" *Aero Magazine*. September 26, 2022. https://physics.nyu.edu/sokal/bad-faith_use_of_words_v2.pdf.

Sokal, Alan. "Transgressing the Boundaries: Towards a Transformative Hermeneutics of Quantum Gravity." *Social Text* 46/47, Science Wars (spring/summer 1996): 217–252.

Sokal, Alan. "Reply to Fish's NYT Op-Ed." Sent to *The New York Times* (unpublished). May, 1996 http://jwalsh.net/projects/sokal/articles/skl2fish.html.

Sokal, Alan. "What is Science and Why Does it Matter? With Professor Alan Sokal." Talk before the Free Speech Union. March 27, 2024. *The Free Speech Union*. Uploaded April 4, 2024. YouTube, 24:01, 55:54 and 1:09:26. https://www.youtube.com/watch?v=UbRrP8UzvSc.

Sokal, Alan. "What the Social Text Affair Does and Does Not Prove." In *A House Built On Sand: Exposing Postmodern Myths About Science*. Edited by Noretta Koertge. 1997.

Sokal, Alan. *Beyond The Hoax: Science, Philosophy and Culture*. 2008.

Solomonoff, Ray. "A Formal Theory of Inductive Inference." *Information and Control* 7 no. 1 (1964): 1–22.

Spencer, Richard. "Richard Spencer's Full Q&A at Auburn University." *The Auburn Plainsman Online (ThePlainsman.com)*. Uploaded 19 April, 2017. YouTube, 14:08. https://youtu.be/g1JJA6UiEio.

Spengler, Oswald. *Decline of the West*. Translated by Charles Francis. 1961.

Starhawk. *Dreaming the Dark: Magic, Sex, and Politics*. 1982.

Starhawk. *The Spiral Dance: a Rebirth of the Ancient Religion of the Great Goddess*. 1979.

Staudenmaier, Michael. "Anti-Semitism, Islamophobia, and the Three Way Fight." *Upping The Anti* 5 (2009).

Sterckx, Roel. "Animal Classification In Ancient China." *East Asian Science, Technology, and Medicine* 23 (2005): 26–53.

Stock, Kathleen. *Material Girls: Why Reality Matters for Feminism*. 2021.

Stock, Kathleen. "The perils of reproductive extremism" *UnHerd*. June 23, 2023. https://unherd.com/2023/06/the-perils-of-reproductive-extremism/.

Stolzenberg, Gabriel. "Review: Kinder, Gentler Science Wars" *Social Studies of Science* 34, no. 1 (Feb., 2004): 77–89.

Sullivan, Alice and Suissa, Judith. "The Gender Wars, Academic Freedom and Education." *Journal of Philosophy of Education* 55, no. 1 (2021): 55–82.

Suppe, Frederick. "Understanding Scientific Theories: An Assessment of Developments, 1969-1998." *Philosophy of Science* 67, Supplement. Proceedings of the 1998 Biennial Meetings of the Philosophy of Science Association. Part II: Symposia Papers (2000): 102–115.

Szilard, Leo. *The Voice of the Dolphins and Other Stories*. 1961.

T. Pennock, Robert. "The Postmodern Sin of Intelligent Design Creationism." *Science & Education* 19 (2010): 757–778.

Tagonist, Anne. "Fuck You Reloaded: Fuck You and Fuck Your Fucking Thesis." *Livejournal*. December 10, 2009. https://tagonist.livejournal.com/199563.html.

Taylor, Chloë. "Foucault, Feminism, and Sex Crimes." *Hypatia* 24, no. 4 (Fall 2009): 1–25.

the Mary Baker Eddy Library. *What did Eddy say about Eastern thought systems?* September 18, 2023. https://www.marybakereddylibrary.org/research/what-did-eddy-say-about-eastern-thought-systems/.

The Out of Order Order. *Liber AAA — The Art of Anarchic Artha: A look through the void via Alan Watts*. 1993. https://www.uncarved.org/OOO/watts.html.

Theocharis, T., Psimopoulos, M. "Where science has gone wrong." *Nature* 332 (June, 1988): 389.

Thomas S. Kuhn. *Objectivity, Value Judgment, and Theory Choice*. 1972.

Thompson, J. M. "Post-Modernism." *The Hibbert Journal* XII, no. 4 (1914): 733.

Thoreau, Henry David. *Walden; or, Life in the Woods*. 1854.

Táíwò, Olúfemi. "Being-in-the-Room Privilege: Elite Capture and Epistemic Deference" *The Philosopher*. October 30, 2020. https://www.thephilosopher1923.org/post/being-in-the-room-privilege-elite-capture-and-epistemic-deference.

van Fraassen, Bas. "The Semantic Approach to Science, After 50 Years." Princeton University and San Francisco State University, April 4, 2014. *Rotman Institute of Philosophy*. Uploaded April 4, 2014. YouTube, 3:53. https://www.youtube.com/watch?v=6oM7-Wa_tAs.

Varma, Roli. "People's Science Movements and Science Wars?" *Economic and Political Weekly* 36, no. 52 (Dec. 29, 2001 – Jan. 4, 2002): 4796–4802.

Vrahimis, Andreas. "Was There a Sun Before Men Existed?" A.J. Ayer and French Philosophy in the Fifties" *Journal for the History of Analytic Philosophy* 1, no 9 (2012) 1–25.

W. Bridgman, P. *Reflections of a Physicist*. 1955.

Wark, McKenzie. "Physicist Opens Fire in the Science Wars" *The Australian*. May 25, 1996. https://web.archive.org/web/20050215055021/http://www.drizzle.com/~jwalsh/sokal/articles/mwark.html.

Watson, David. *Against The Megamachine: Essays on Empire and It's Enemies*. 1998.

Weinberg, Steven. "Conceptual foundations of the unified theory of weak and electromagnetic interactions." *Reviews of Modern Physics* 52, no. 3 (1980): 515–523.

Weinberg, Steven. "Peace at Last?" In *The One Culture?* Edited Jay A. Labinger and Harry Collins. 2001.

Weinberg, Steven. "The Revolution That Didn't Happen" *New York Review of Books*. October 8, 1998. https://www.nybooks.com/articles/1998/10/08/the-revolution-that-didnt-happen/.

Weinberg, Steven. *Dreams Of A Final Theory: The Search for The Fundamental Laws of Nature*. 1993.

Weinstein, Brett (@BrettWeinstein). Twitter. "'Civility is racist' was a central principle of the Evergreen revolutionaries and their faculty mentors, even if they never said it in exactly those words." Twitter, April 26, 2024. https://archive.is/NCgMJ.

Werbre, Peter. "Decades in the Struggle - Interview with Peter Werbe." June 30, 2025. *Nathan Jun*. YouTube, 14:01 https://www.youtube.com/watch?v=yoYVhMd55hI.

Werskey, Gary. *The Visible College: A collective biography of British scientists and socialists of the 1930s.* 1978.
West, Patrick. "In defence of postmodernism" *Spiked.* April 15, 2023. https://www.spiked-online.com/2023/04/15/in-defence-of-postmodernism/.
Wilkins, John S. "How not to Feyerabend." *Evolving Thoughts.* October 5, 2007. https://evolvingthoughts.net/how_not_to_feyerabend/.
Williams, Jeffrey J. "Actually Existing Cosmopolitanism" *LA Review of Books*, December 23, 2018. https://lareviewofbooks.org/article/actually-existing-cosmopolitanism/.
Wilson, Peter Lamborn. *Boundary Violations* in Technoscience and Cyberculture. Edited by Stanley Aaronwitz. 1996.
Wilson, Peter Lamborn. *Millennium.* 1996.
Wilson, Robert Anton. *Cosmic Trigger III.* 1995.
Wolfson, Louis. "Full Stop for an Infernal Planet or The Schizophrenic Sensorial Epileptic and Foreign Languages." In *Hatred Of Capitalism: A Semiotext(e) Reader.* Edited by Chris Kraus and Sylvère Lotringer. 2001.
Wolin, Richard. ""The Leprosy of the Soul in Our Time": On the European Origins of the "Great Replacement" Theory" *Los Angeles Review of Books*, August 5 2022. https://lareviewofbooks.org/article/the-leprosy-of-the-soul-in-our-time-on-the-european-origins-of-the-great-replacement-theory/.
Wolin, Richard. *Heidegger In Ruins: Between Philosophy and Ideology.* 2022.
Wong, Mia (@itmechr3). "An academic will help the cops arrest you while sneering that if you don't think a rock is a person you're reinscribing western colonialism." Twitter, 2022. URL lost.
Woods, Juidth. "Kathleen Stock: 'No matter what I say, to trans people I'll always be a villain'" *The Telegraph.* May 28, 2023. https://www.telegraph.co.uk/news/2023/05/28/kathleen-stock-interview-oxford-university-gender-debate/.
Woolgar, Steven. *Knowledge and Reflexivity.* 1988.
Woolgar, Steven. *Science: The Very Idea.* 1988.
Wylie, Alison. "The Interplay of Evidential Constraints and Political Interests" *American Antiquity* 57, no. 1 (Jan 1992): 15–35.
Zarlengo, Kristina. "Idiot Savants?" *Salon,* November 1998. https://www.salon.com/1998/11/02/cov_02feature/.
Zerzan, John. "Against Technology: A talk by John Zerzan." Lecture at the Discourse@Networks 200 symposium held at Stanford University. April 23, 1997. https://theanarchistlibrary.org/library/john-zerzan-against-technology-a-talk-by-john-zerzan-april-23-1997.
Zerzan, John. *The Catastrophe of Postmodernism.* 1991.
Žižek, Slavoj. *Less Than Nothing: Hegel and The Shadow of Dialectical Materialism.* 2012.

ABOUT THE AUTHOR

William Gillis was born into the anarchist movement and has organized in varied capacities since the Battle of Seattle in 1999. Gillis' writing on diverse subjects can be found on HumanIterations.net. Other works include *Science As Radicalism*, and the forthcoming compilations *Bookshelf Engagements* and *Zine Crate Provocations*.

www.ingramcontent.com/pod-product-compliance
Lightning Source LLC
Chambersburg PA
CBHW031308150426
43191CB00005B/120